Dr. Jerry E. Sipe
Anderson College
November, 1974

WALTER E. HARRIS

Professor of Analytical Chemistry
University of Alberta

BYRON KRATOCHVIL

Professor of Analytical Chemistry
University of Alberta

CHEMICAL SEPARATIONS AND MEASUREMENTS: Background and Procedures for Modern Analysis

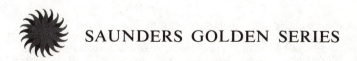

SAUNDERS GOLDEN SERIES

W. B. SAUNDERS COMPANY
Philadelphia • London • Toronto • 1974

W. B. Saunders Company: West Washington Square
Philadelphia, PA 19105

12 Dyott Street
London WC1A 1DB

833 Oxford Street
Toronto, Ontario M8Z 5T9, Canada

Chemical Separations and Measurements: ISBN   0-7216-4535-6
An Introduction to Modern Analytical Measurements

Last digit is the print number: 9   8   7   6   5   4   3   2   1

*When it has once been given a man to
do some sensible things, afterwards his
life is a little different.*

A. Einstein

PREFACE

This book is an expansion and major revision of our earlier edition, *Chemical Analysis*, by another publisher. In that edition we began the preface with a statement that is appropriate to repeat here.

> "We believe that the contents of an introductory course in analytical chemistry should be related to the activities in modern chemical laboratories. Chemists are involved primarily with gathering and interpreting information about chemical systems. Thus, sound training for a chemist consists in developing an ability to perform these two functions. The aim of this book is to provide a foundation for the first of these functions, that of gathering reliable experimental data."

The trend toward teaching a combination of noninstrumental and instrumental topics in the introductory analytical course, mentioned in that preface, is now well established. This book is designed to provide a set of tested experiments that bring the laboratory treatment up to the level of the best discussions of theory and principles. No prior analytical experience is assumed; the experiments begin with basic analytical operations and build progressively to include current techniques such as gas chromatography and atomic absorption. The level of the material is designed for students who have a background of general chemistry and who have gained sufficient appreciation of the need for high-quality experimental work to benefit from an intensive course in chemical measurements. The goal is to build the confidence of the student in his ability to make accurate chemical measurements at a reasonable rate. For students specializing in chemistry, this course should be followed by one in chemical instrumentation.

The first chapter introduces the basic laboratory tools and operations; this is followed by a brief treatment of handling of data. Chapters 3 through 5 treat noninstrumental methods; 6 through 9, instrumental methods; and 10 and 11, separations. The final chapter introduces the concept of analytical problem solving and the use of chemical literature. These 12 chapters contain more than enough experiments for a one-year course, so a choice is available in most areas.

The detailed instructions for the procedures have been chosen, tested, and refined over a period of years. Unknown samples are used throughout; there is objective evidence that they increase both teaching effectiveness and student interest. At Alberta, from 8 to 10 hours of laboratory time is spent on the experimental work in Chapters 1 and 2, from 45 to 55 hours on work in Chapters 3, 4, and 5, and from 50 to 65 hours on work selected from Chapters 6 through 12. Data collected on the median time required by students to complete the experimental work are given for most experiments to aid in planning and selection and to help students in planning efficient use of their time. Our laboratory is accompanied by about 75 hours of lectures on the theory and practice of analytical chemistry, which are supplemented by a modern introductory analytical text. The companion text to this laboratory book, by Peters, Hayes, and Hieftje, is recommended.

We thank Dennis Peters of the University of Indiana for reading the entire manuscript. He also provided (from the 3rd edition of Quantitative Chemical Analysis) Experiment 10-2 on the ion exchange separation of zinc and nickel, Experiment 11-3 on determination of nickel in copper by extraction as the DMG complex, and several photographs.

We are indebted to Dorothy Cox for helpful discussions and ideas during the preparation of this edition; to our teaching assistants of the past twelve years for their comments, advice, and questions; to Phyllis Harris for her invaluable help in matters of consistency and clarity; to Glen Johanson for preparation of the illustrations; and to the many others who have contributed.

W. E. HARRIS
B. KRATOCHVIL

CONTENTS

Chapter 1

PRELIMINARY EXERCISES AND INFORMATION

If from any art you take away that which concerns weighing, measuring, and arithmetic, how little is left of that art!

Plato

This chapter provides an introduction to the basic equipment and techniques used in succeeding chapters. It includes (1) general introduction to laboratory work, (2) weighing, (3) correct use of volumetric glassware, (4) treatment and evaluation of experimental data, and (5) volumetric calculations.

1–1 LABORATORY SAFETY AND CLEANLINESS

General safety practices should be observed at all times. Although this text contains no specifically hazardous experiments, danger is inherent in all chemical experimentation. Carelessness leads to accidents, and warnings cannot replace common sense and reasonable caution. A list of basic precautions follows:

1. Learn the location of the safety shower and when it should be used.
2. Report every accident to the laboratory instructor. First aid will be provided. If necessary you will be taken for emergency treatment; do not try to go alone.
3. Wear eye protection in the laboratory. Ordinary prescription glasses are sufficient, or a pair of inexpensive safety glasses. Even in a procedure as innocuous as washing glassware, protection may be needed from flying glass or from chemicals inadvertently splashed by nearby workers.
4. Protect both hands with several thicknesses of cloth toweling while inserting glass tubing into rubber stoppers. First wet the tubing with

1

soapy water or dilute aqueous ammonia, and then apply pressure slowly with a twisting motion.

5. Beware of burns from forgotten, still lit burners and from hot glass.
6. Organic solvents are generally flammable. Do not use them near an open flame.
7. Some chemicals, such as mercury(II) chloride and sodium cyanide, are highly poisonous. Handle these materials with care; wipe up spilled material promptly and completely.

Keep all equipment and glassware in good condition. Glassware should be cleaned with a brush and detergent as soon as possible after use. Not only is it easier to clean, but it is then always ready for use, and possible contamination of reagents and samples is avoided. Volumetric glassware must be sufficiently free of grease to drain smoothly and uniformly without drops of solution adhering to the interior walls. Burets should be stored filled with distilled water. At the end of each laboratory period clean and put in order your work space and any community equipment or space you used.

1–2 PRELIMINARY OPERATIONS

This section lists background exercises and directions for preparation of equipment for the experiments that follow. Your instructor will indicate any changes necessary.

1. Obtain a set of volumetric equipment (if not provided in your locker): a buret, a 10-ml pipet, a 20-ml pipet, six 50-ml volumetric flasks, two 100-ml volumetric flasks, one 250-ml volumetric flask, and one 500-ml volumetric flask. Calibrate the pipets, flasks, and buret as described in Section 1-5.
2. Carry out the balance assignment described in Section 1-4. Record observations and conclusions in a laboratory notebook (see Section 1-3).
3. Check in locker equipment. Examine particularly the tips of the buret and pipets for chips that may affect reproducibility of delivery. If in doubt about the acceptability of a piece of equipment, consult the instructor.
4. Use a calculator to solve the problems listed in Section 2-1. Check all answers for the correct number of significant figures.
5. Do the aliquoting exercise outlined in Section 1-5.
6. Read Section 2-1 on evaluation of experimental data, and solve the problems at the end of the section.
7. Prepare or obtain a 1-liter polyethylene wash bottle with the tip

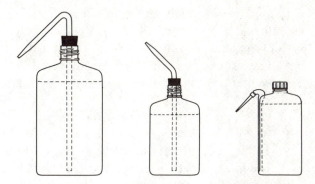

FIGURE 1–1.   Three types of polyethylene wash bottles.

pointing downward. In addition, prepare a 500-ml polyethylene wash bottle with the tip pointing upward. (See Figure 1-1 for illustrations.) An upward-pointing tip is useful in the quantitative transfer of solutions and solids. When tips are prepared in the laboratory, always place hot glass on wire gauze, not directly on the bench top. Protect your hands (Section 1-1). Use a stopper with a hole large enough to accept the glass tubing without forcing.

8. Examine the ordinary crucibles for individual distinguishing marks. If necessary, mark with small dots of 10% cobalt(II) chloride solution on the outer surface near the top, and heat strongly for 10 to 15 min over a Bunsen burner. Cool, and then scrub the markings with a brush and water to remove cobalt(II) chloride that did not fuse into the porcelain.

9. Prepare three or four stirring rods 20 to 25 cm long from 5-mm glass rod. Fire-polish both ends of each rod.

10. Read Section 1-3 and the experimental-procedure section of Experiment 3-1, and then prepare a blank summary data page. Ask the laboratory instructor to check the prepared data page and initial it.

11. If a practical test is scheduled, review Items 1, 2, 4, 5, and 6. The test may consist of items such as a short written quiz, reading several burets, using a calculator to multiply and divide, weighing with an analytical balance, and taking an aliquot of a solution.

1–3   THE LABORATORY NOTEBOOK

A laboratory notebook is designed to provide a permanent record of firsthand observations. The criterion for sound record keeping is that someone else can readily locate pertinent data and results for an experiment. Although reasonable legibility and neatness are desirable, the usefulness of a record is determined largely by whether it is original, systematic, and complete—not by whether it is a work of art.

The notebook should be hard-covered and bound. A 7- by 9-in. size is convenient. Number the pages consecutively, leaving two or three pages at the front for an index.

To keep a proper experimental record:

1. Enter data directly in ink as soon as taken. Never recopy numbers or use loose sheets of paper. Cancel errors or rejected data by drawing a single line through them. Do not erase or remove pages. The notebook should be a permanent record of the original laboratory work.
2. Enter data only on the right side of the page. Use the left side for recording observations that may be useful in evaluating results, for calculations, and so on.
3. Clearly label all entries. To facilitate direct entry of experimental work, set up a data page for each experiment before starting. Examples are shown below.
4. Make the record of experiments complete but concise. Date each entry. Give each experiment a title and list it in the index.

EXAMPLES OF SUMMARY DATA PAGES

Standardization of HCl against Na_2CO_3	Date *October 10, 1973*			
	1	**2**	**3**	**4**
Wt weighing bottle + sample, g	19.2750	18.8346	18.4640	18.0229
Wt weighing bottle, g	18.8346	18.4640	18.0229	17.6317
Wt Na_2CO_3 sample, g	0.4404	0.3706	0.4411	0.3912
Final buret reading	40.22		43.41	40.97
Initial buret reading	0.14	1.03	3.23	3.48
Difference	40.08	End point missed	40.18	37.49
Net buret correction	+0.02		+0.01	+0.02
Volume HCl (corrected)	40.10		40.19	37.51
Molarity of HCl	0.2072		0.2073	0.2069
Average molarity of HCl		0.2071		
Relative average deviation		0.8 ppt		
Median		0.2072		
Relative probable deviation		0.5 ppt		

Determination of Na$_2$CO$_3$ in a Sample (Sample Code Number B-231)	Date October 10, 1973			
	1	2	3	4
Wt weighing bottle + sample, g	16.4361	16.0632	15.6321	15.2207
Wt weighing bottle, g	16.0632	15.6321	15.2207	14.8100
Wt sample, g	0.3729	0.4311	0.4114	0.4107
Final buret reading	33.96	43.90	38.31	36.44
Initial buret reading	1.23	5.84	1.86	0.16
Difference	32.73	38.06	36.45	36.28
Net buret correction	+0.04	+0.01	+0.03	+0.03
Volume HCl (corrected)	32.77	38.07	36.48	36.31
% Na$_2$CO$_3$ (from average HCl M)	96.45	96.93	97.34	97.02
% Na$_2$CO$_3$ (from median HCl M)	96.50	96.98	97.38	97.07
Average % Na$_2$CO$_3$	96.94%			
Relative average deviation	2.3 ppt			
Median % Na$_2$CO$_3$	97.02%			
Relative probable deviation	2.0 ppt			

1–4   WEIGHING

> If you can measure that of which you speak, and can express it by a number, you know something of your subject; but if you cannot measure it, your knowledge is meager and unsatisfactory.
>
> *Lord Kelvin*

Weighing provides the foundation on which exact chemical knowledge rests. The precision with which analytical measurements can be made depends ultimately on proper operation of the correct kind of balance. The quality of experimental work in this course therefore depends on the balance and how it is used. In this section the design and operation of the analytical balance, the top-loading balance, and the triple-beam balance are described.

The Analytical Balance: Background

The user of an analytical balance should know how to determine whether it is operating properly and how to make the simple adjustments

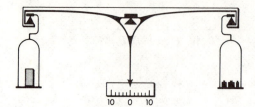

FIGURE 1–2. Schematic illustration of the null-balance principle in weighing with a two-pan balance. Weights are added to opposite arm of lever until beam returns to zero position, as indicated by attached pointer. Load on beam varies with weight of object.

that are occasionally necessary. After instruction, adjustments in optical-scale sensitivity and zero position of a single-pan balance can be made easily in 5 to 10 min.

Weighing usually involves application of the lever principle. The weight[1] of an object is determined by (1) addition or removal of weights from one arm of the lever until it is brought back to its initial position (the null-balance principle, illustrated in Figure 1-2) or (2) measurement of the extent to which the lever is deflected from its initial position by the weight of the object (the direct-deflection principle, Figure 1-3). Most often used is a combination of null balance and direct deflection in which the direct-deflection principle is employed for the last fraction of the weight.

Until recent years almost all analytical balances were of the two-pan type. They are highly precise but tedious to operate. They also are subject to decreasing sensitivity with load, that is, the larger the load on the beam, the smaller the beam deflection per unit of weight. This variation in sensitivity makes the use of the direct-deflection principle in two-pan balances unreliable for the determination of any but a minute fraction, typically 1 to 2 mg, of the weight of an object.

In the constant-load type of single-pan balance a large counterweight is permanently attached to one end of the beam, making a knife-edge unnecessary at this end (Figures 1-3 and 1-4). The analytical weights and the pan holding objects to be weighed are suspended by a knife-edge from the opposite end. The object to be weighed is placed on the pan, and

---

[1] The term *weight* rather than *mass* is used throughout this book.

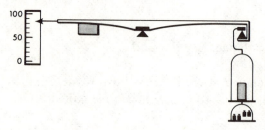

FIGURE 1–3. Schematic illustration of the direct-deflection and constant-load principles in a single-pan balance. Weights are removed from lever arm supporting object until beam returns to near zero position. Last portion of weight is read from optical scale. Beam is under constant load at all times.

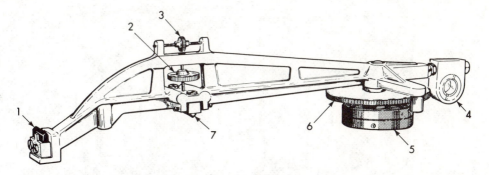

FIGURE 1—4. Beam from a constant-load balance: (1) terminal knife-edge; (2) nut for adjustment of optical-scale sensitivity; (3) coarse zero-adjustment knob; (4) reticle for optical-scale readout; (5) counterweight; (6) piston portion of air damper and part of counterweight; (7) central knife-edge.

weights are removed until they correspond to slightly less than the weight of the object. The last fraction of the weight of the object is obtained by use of the direct-deflection principle and is read from an optical scale. Because the beam has a constant load upon it, the sensitivity of the balance is independent of the weight of the object and a large fraction of the weight may be obtained by direct deflection. The full optical scale is typically either 100 mg or 1 g in a balance of 100 to 200-g capacity. For a 100-mg scale the deflection is read from an image of a 0- to 100-mg scale attached to the beam and projected onto a frosted glass or plastic panel near the front of the balance. This optical scale can be read to the nearest 0.1 mg. The use of direct deflection for the last fraction of the weight, along with manipulation of the larger weights by exterior knobs, makes weighing much more rapid and convenient with a single-pan balance than with a classical two-pan balance. An additional advantage of the constant-load balance is that accuracy and precision can be maintained over long periods of time because only two knife-edges are required instead of three and because the use of internal weights minimizes the chances of weight changes from corrosion or mishandling.

It is important that an analytical balance be treated with the care due any precision instrument. It should be mounted on a base that minimizes the effects of vibration. To reduce wear on the knife-edges, arrest the beam whenever a reading is not being taken. Check the zero setting frequently and the optical scale periodically. Each should be adjusted as needed according to the procedure outlined in Items 5 and 6 in the experimental work that follows. Two typical analytical balances are shown in Figure 1-5. A schematic drawing of the internal components is shown in Figure 1-6, and the external controls of a different model in Figure 1-7.

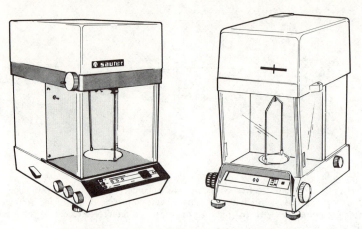

FIGURE 1–5. Two single-pan analytical balances.

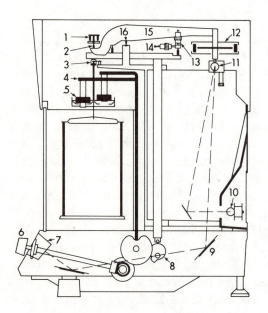

FIGURE 1–6. Schematic drawing of components of a single-pan analytical balance: (1) pan-support stirrup; (2) front knife-edge; (3) pan brake; (4) weight carriage; (5) built-in weights; (6) weight control mechanism; (7) optical readout; (8) beam-arrest shaft; (9) mirror (adjustable for zero-point setting); (10) bulb; (11) scale and objective; (12) damping; (13) sensitivity adjust; (14) coarse-zero adjust; (15) beam; (16) center knife-edge.

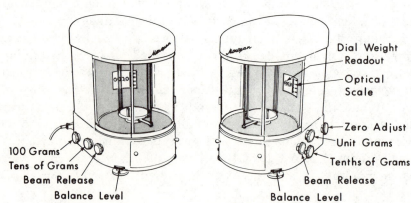

FIGURE 1–7. External components of a single-pan analytical balance.

The Analytical Balance: Experimental

Operating Instructions

Before every weighing, see that the balance is level, the pan compartment clean, and the door to the compartment closed. Turn the beam-arrest knob to full release, and with the fine-zero knob adjust the balance to read zero. Arrest the beam, put the object to be weighed on the pan, and carefully turn the beam-arrest knob to the partial-release position.[2] Remove weights by first rotating the knob for tens of grams so that the numbers appear in the order 9, 8, 7, 6, . . . until the lighted portion of the optical scale changes from a position less than zero to one greater than zero. Next, remove smaller weights by rotating the knob for grams. Finally, rotate the knob for tenths of a gram (if applicable) until the optical scale reads greater than zero. Then carefully turn the beam-arrest knob to full release. Read the weight, using the vernier for tenths of milligrams. Arrest the beam, remove the object, dial the weights to zero, and check the zero reading. If it has changed, reset the zero and reweigh the object.

When finished, leave the balance clean and in the arrest position. Dial the weights to zero, and close the door to the pan compartment. Never put chemicals directly on the pan; use a container such as a beaker or a flask. Wipe up any spilled material immediately.

Experimental Work

Before beginning work, read the background information on the balance and obtain instruction on its design and correct operation. Review the technique for reading a vernier scale, if applicable. For most measurements, greatest precision is achieved by an edge-to-edge technique as in Figure 1-8A (see also Figure 9-4); a vernier scale, however, is more

---

[2] Some balances incorporate a *preweigh* feature. In these balances the partial-release position is replaced by a preweigh position. The object to be weighed is placed on the pan and the beam-arrest knob turned to preweigh. The weight of the object is read to the nearest gram from the optical scale; this weight is dialed, the beam-arrest knob turned to full-release position, and the final weight read in the usual way.

FIGURE 1–8. Correct methods for reading scales: (A) meters, slide rules, rulers, and so forth; (B) verniers.

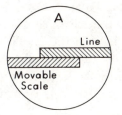

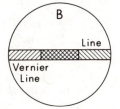

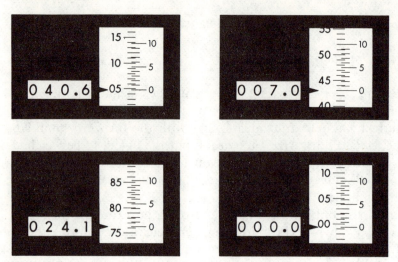

FIGURE 1–9. Diagrams for practice in vernier reading. (See Appendix D for correct readings.)

conveniently read at the position where the lines superimpose, as in Figure 1-8B. Practice on the examples in Figure 1-9.

    A balance will be assigned for your use during the course. After an instruction session, complete the following operations on that balance.

1.   **Reproducibility of Balance Readings with and without Load.** Zero the balance by releasing the beam and adjusting the fine-zero knob. Arrest the beam. With forceps add a crude 100-g weight to the pan. Carefully dial the appropriate knob to remove 100 g of balance weights, and release the beam. Record the reading. Arrest the beam, remove the weight, dial zero grams, and record the new no-load reading. Do not reset the zero. Reload the 100-g weight onto the pan and record the reading; then remove the 100-g load and again determine the no-load reading. Repeat these operations three or four times. Calculate the average no-load reading, the average reading for the 100-g weight (taking into consideration any change in the preceding no-load reading), and the average deviation for each of these readings.

2.   **Test of Optical-Scale Sensitivity.** Add a crude 100-mg weight (or 1-g weight if the balance has a 1-g optical scale) to the pan and dial 0.1 g. Release the beam and zero the optical scale with the zero adjustment. Now dial 0.0 g. (This step adds the 0.1-g balance weight to the front end of the beam.) Record the reading of the optical scale. Repeat the above operations to obtain a duplicate reading of the optical scale. Does the optical scale read correctly? Must it be in

correct adjustment for the tests of sensitivity and internal consistency in Items 3 and 4 below?

3.   **Sensitivity as a Function of Load.** Add crude 100-mg and 100-g weights to the pan. Dial 100.1 g. Release the beam and set the optical scale to read zero. Now dial 100.0 g and read the optical scale. Arrest the beam. Remove the 100-g weight and dial 0.1 g. Set the optical scale to read zero. Now dial 0.0 g and read the optical scale. Repeat the above operations.

The difference between the optical-scale deflection with and without the 100-g load is a measure of sensitivity error. Does the sensitivity vary with load (sensitivity error)? Since the balance operates under constant load, should the optical-scale reading be affected by the load on the pan?

4.   **Internal Consistency of Weights**. Add a crude 100-g weight to the pan, dial 100.0 g, release the beam, and record the reading. Arrest the beam. Now dial (carefully) 99.9 g, release the beam, and again record the reading. Repeat these operations. Are the weights internally consistent? If they are, the difference in the optical-scale readings should be the same as the difference observed in Item 2.

Compensating errors are not detected by the above operations. Generally speaking, the weights built into constant-load balances are internally consistent; that is, a large weight is the sum of smaller weights plus the full optical-scale reading.

Before proceeding to Items 5 and 6, ask the laboratory instructor to put the balance significantly out of zero and sensitivity adjustment.

5.   **Adjustment of Balance Zero.** Release the beam. Note that the balance cannot be set at zero with the fine-zero knob. Arrest the beam, remove the balance cover, and give the coarse-zero nut one or two turns. Replace the cover, release the beam, and observe the deflection of the optical scale. Continue as above until the balance reads zero with the fine-zero knob approximately centered.

6.   **Adjustment of Optical-Scale Sensitivity.** Adjust the optical scale to be internally consistent with the balance weights as follows:

a.   Add a crude 100-mg weight to the pan (or 1 g if optical scale is 1 g).

b.   Dial 0.1 g. Release the beam and set the optical scale to read zero.[3] Dial the balance back to 0.0 g and read the optical scale.

---

[3] Setting the optical scale to read zero usually will displace it from the original zero of the balance because the crude weight is not exactly 100 mg. This displacement is unimportant, as the only purpose of this weight is to allow the balance to be zeroed with the 0.1-g weight dialed. In this way the optical scale can be checked against the 0.1-g internal weight simply by dialing 0.0 and observing the optical-scale reading. Using a crude 100-mg weight along with setting the optical scale to zero with 0.1 g on the dial eliminates the need for a high-quality external 100-mg weight by substituting for it the one in the balance.

c. Arrest the beam and remove the cover. Turn the sensitivity nut in the proper direction and replace the cover. Repeat Step b. Continue until the optical scale reads correctly within 0.1 mg. The scale is now consistent with the 0.1-g internal balance weight. The optical-scale adjustment should be checked every two or three weeks and adjustments made as necessary.

7. **Weighing.**

a. The responsiveness of a balance to small differences in weight may be illustrated as follows. Weigh a 50-ml beaker containing a few milliliters of water. Determine the rate of evaporation of the water by observing the reading at 1-min intervals for 5 min. What is the rate of evaporation in milligrams per minute?

b. Weigh an object supplied by the laboratory instructor. Report to him the value obtained.

Comments on Experimental Work

When a balance is operating properly, the beam-release mechanism operates smoothly, the beam comes to a stable rest point in a few seconds, the zero point is the same before and after a weighing, and consecutive readings for an object are the same. A test of the reproducibility of the readings with and without load gives an impression of what to expect of a balance and a justified feeling (or lack) of confidence in its performance.

A group of people may increase the temperature of a previously empty balance room by several degrees; this rise in temperature may cause the zero point of a balance to shift as much as a milligram within a few minutes. In such a situation the zero point should be observed frequently.

For a properly designed balance in which part of the weight is obtained by the direct-deflection principle, the sensitivity must not change with load on the pan. In the test of sensitivity in Item 3 a constant-load balance should give the same optical-scale readings with and without load. To pass this test, the optical scale need not be in correct adjustment; the only requirement is that it read the same with and without a load on the pan.

The internal-consistency check can be carried out at levels other than 100 g such as 5, 10, or 50. It gives an idea of the quality of the weights, but because compensating errors can occur, it is in no sense a calibration of the weights. Remember also that discordant results may be obtained if the optical scale is not in adjustment. The optical scale will read full scale only if the weights are internally consistent and if the optical scale is correctly adjusted. Because the quality of the weights supplied with constant-load balances is usually equal or superior to Class S tolerances,[4] weight calibrations of the bad old days are unnecessary.

The procedure for optical-scale adjustment outlined here is strongly recommended over the more obvious one of calibration against an external standard weight. First, the internal 100-mg weight supplied with the balance is of high quality and, because it is not handled, remains so. Second, the adjustment of the optical scale can be completed more quickly and with less confusion because the crude 100-mg weight is left on the pan throughout the operations. The procedure is not only faster, but more foolproof, particularly against failure to realize that turning the sensitivity nut usually results in a change of the zero.

Routine Weighing on an Analytical Balance

Selection of the best method for weighing depends on the number and nature of the samples. When only one or two samples are required, *direct weighing* is preferable. To carry out this procedure, first weigh an empty vessel and record the value. Set the balance to read an additional amount equal to the weight of sample to be obtained. Add sample in small amounts until approximate balance is achieved. Again weigh and record. The difference in readings is the sample weight. With this technique each sample requires two weighings. Sample may be added while the vessel is on the balance pan only if the sample is not corrosive and no material is spilled in the weighing compartment.

When several samples are required, *weighing by difference* is sometimes employed. To use this technique, put sufficient material for all the samples into a vessel. Weigh the vessel and contents and record the reading. Set the balance to read less by the approximate weight of sample desired. Using a camel's-hair brush and a spatula, quantitatively transfer small amounts of material to a clean beaker or flask until the approximate weight is reached. Weigh the vessel precisely again. The difference between the two weighings is the weight of the first sample. The weight of the second sample is obtained in the same manner as that of the first, but since the initial weight is already known, only one additional weighing is needed. Thus three samples require only four weighings.

Some balances are equipped with a built-in compensator (taring mechanism) that enables the operator to set the balance to zero with an empty vessel on the pan. In this way a sample can be weighed directly into

---

[4] The tolerances for various classes of weights are defined by the National Bureau of Standards, Washington, D.C., *Circular 500*. Class S tolerances for 100-mg, 1-g, and 10-g weights are 0.025 mg, 0.054 mg, and 0.074 mg.

a beaker or flask with only one weighing (excluding the zero reading) per sample. This is the most efficient method of all. The principal advantage of weighing by difference is that it enables hygroscopic or volatile samples to be weighed with little exposure to the atmosphere.

For accurate weighing of corrosive materials, approximate amounts are weighed first on a triple-beam or other auxiliary balance to protect the analytical balance from spillage and corrosive vapors. The procedure for weighing silver nitrate or iodine, for instance, is as follows. Weigh an empty vessel on an analytical balance (using a cover if the material is volatile). Record the weight. Take the vessel to a rough balance and weigh into it the approximate amount of material to be taken. Accurately reweigh the vessel plus material on the analytical balance. The vessel must not be handled directly between analytical weighings; use tongs, finger-cots, or other protection.

Potential sources of error in weighing with a constant-load balance include inaccurate weights, shifts in balance zero or sensitivity, changes in the sample, and air buoyancy. The first two have been discussed. Errors from weight changes in the sample include absorption of moisture (especially serious when a hygroscopic material is being weighed), volatilization, air currents from hot sample or container, and static charge on containers. Errors from air buoyancy occur because an object immersed in a gas or a liquid is buoyed up by a force equal to the volume of fluid displaced. This gives the weight of an object weighed in air an apparent weight that is less than the true weight by the weight of the displaced air. The same effect, of course, operates as well on the weights used. If the density of the weights and of the object are equal or nearly so, buoyancy errors are zero or negligible; stainless-steel weights, the most common type in current use, have a density of 7.8. With such weights a correction for buoyancy is necessary at the level of part-per-thousand accuracy only when the density of the object being weighed is 1.0 or less. This is encountered primarily when liquids are weighed. The buoyant effect of air on an object B_{obj} is given by

$$B_{obj} = \frac{\text{apparent weight}}{\text{density of object}} \times \text{density of air} \qquad (1-1)$$

The weight of an object divided by its density gives the volume of the object and therefore the volume of air displaced. Multiplying this volume by the density of air[5] gives the weight of air displaced and thus the

---

[5] The density of dry air depends on barometric pressure and temperature. At 25°C a density of 0.0011 g/ml can be used for barometric pressures in the range 670 to 740 mm of mercury, and 0.0012 g/ml for pressures above 740 mm. A pressure of 740 mm corresponds to an altitude of about 300 m. At a pressure of 760 mm, 0.0012 g/ml can be used over the temperature range 10 to 33°C.

buoyant force on the object. Similarly, the buoyant effect of air on stainless-steel weights is

$$B_{wts} = \frac{\text{apparent weight}}{7.8} \times \text{density of air} \qquad (1-2)$$

The net buoyant effect is the difference between these two values, that is, $B_{obj} - B_{wts}$; this difference is added algebraically to the apparent weight to give the true weight. The use of buoyancy corrections in weighing is illustrated in Example 1-1 in the procedure for pipet calibration.

The Top-Loading Balance

A top-loading balance is used for rapid weighings with a precision between that of a triple-beam balance and an analytical balance. Two top-loading balances are shown in Figure 1-10 and a schematic drawing of the internal components in Figure 1-11. This type of balance operates under constant load with few or no internal weights. The range in optical scale is large; four significant figures normally can be obtained from it alone.

After an object is placed on the pan, the balance comes to rest with the aid of magnetic damping in 3 to 4 sec. Often provided is a taring device that minimizes weighing errors by eliminating the need for recording and subtracting the weight of the sample container. A taring knob controls a small spring that exerts a force on the beam to bring the optical scale back to zero. This permits the weight to be read directly from the optical scale when sample is added to the container. Because the pan of a top-loading balance is not ordinarily enclosed in a case, it must be protected from air currents by some other means. A variety of shields are available. The optical-scale sensitivity of a top-loading balance must be checked periodically in the same way as that of an analytical balance. In those models having no internal weights an accurate external weight must be used for this operation.

 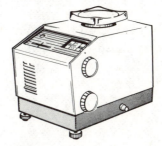

FIGURE 1–10. Two top-loading balances.

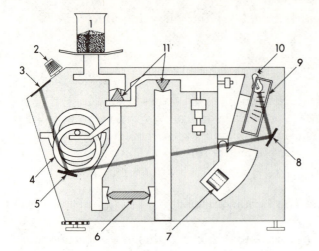

FIGURE 1–11. Schematic drawing of components of a top-loading balance: (1) object weighed; (2) weight knob; (3) optical readout; (4) weights; (5) mirror; (6) guide plate; (7) magnet; (8) mirror; (9) optical scale; (10) lamp; (11) knife-edges.

The capacities of the balances of one manufacturer (Mettler) range from 2 mg to 10 kg. The model with 2-mg capacity can weigh objects to the nearest 0.1 or 0.2 μg, and the 10-kg industrial model to the nearest 0.5 g. In both, the optical scale extends over the full weighing range, and no beam arrest is used.

For semiquantitative laboratory work a top-loading balance with a capacity of about 150 g and a sensitivity of about 1 mg is useful. (The Mettler P-160 model has a capacity of 160 g, an optical-scale range of 10 g, a tare capacity of 10 g, a sensitivity of 1 mg, and a set of weights up to 150 g in 10-g intervals that is operated by an external knob. The beam bearings are of sapphire.)

Operating Instructions for a Top-Loading Balance with
10-g Optical Scale

1.   Ensure that the balance is level.
2.   Set the optical scale to zero.
3.   Place the object to be weighed on the pan. If the weight exceeds 10 g, add weights in 10-g increments as needed by turning the weight knob.
4.   Read and record the weight of the object. (On some balances a knob may be dialed to assist in obtaining the terminal digit.)

The Triple-Beam Balance

Triple-beam balances are designed for rapid weighings with an accuracy of 0.01 to 1 g, depending on the capacity. Three graduated scales, each with a weight attached, are used. The maximum capacity ranges from about 100 g to several kilograms. The beam typically has a zero-adjust knob, two knife-edges of hard metal, and bearings of agate. A beam-arrest mechanism is not usual.

PROBLEMS[6]

1–1.   A balance was tested for variation of sensitivity with load according to Item 3 in the experimental procedure. An optical-scale reading of 97.3 mg was obtained with 100.0 g of weights dialed; an optical-scale reading of 97.2 mg was obtained with 0.0 g dialed. Is there a sensitivity error? (Recall that an analytical balance weighs to a precision of about 0.1 mg.) What corrective action should be taken before the balance is used for weighings?

1–2.   The optical-scale setting of a balance can be tested as follows: Add a rough 100-mg weight to the pan. Dial 0.1 g. Release the beam and set the optical scale to zero. Dial 0.0 g. Read the optical scale. Arrest the beam. The rough weight need not be exactly 100 mg. Why? Why is no sensitivity error expected in a constant-load balance?

1–3.   What is the true weight of an object having an apparent weight of 20.0000 g when weighed with stainless-steel weights if the object is pure aluminum (density = 2.7); platinum (density = 21.4); stainless steel (density = 7.8); borosilicate glass (density = 2.7)?

1–4.   Devise a set of seven individual weights that will provide any combination of weights from 0 to 100 g at 1-g intervals. If one more weight were permitted, what set might be selected?

1–5.   The optical scale of a balance reads 97.5 instead of 100.0. If two revolutions of Nut 2 in Figure 1-4 are required to change the sensitivity 1 mg, in what direction and how much should it be turned for the optical scale to read correctly?

1–6.   A sample of NaCl (density = 2.16) had an apparent weight of 5.0000 g when weighed on a single-pan analytical balance (stainless-steel weights). What is the true weight of the NaCl?

1–7.† When making buoyancy corrections, one *must include* that portion of the weight obtained from the optical-scale reading. Why?

---

[6] The more difficult or challenging problems throughout the book are indicated by a dagger.

1–5 VOLUMETRIC MEASUREMENTS

Do not descend into a well on an old rope.

Turkish proverb

For making accurate measurements in analytical procedures, next in importance to the balance is volumetric equipment. In this section volumetric flasks, pipets, burets, and graduated cylinders are discussed. The experimental procedure includes gravimetric calibration of pipets, burets, and volumetric flasks to provide checks on the accuracy of the markings as well as on proper technique. The few minutes required to learn correct use of this equipment will save time in the long run and give better results. For example, a typical experienced person using incorrect technique takes about 10 sec to read a buret and will have a standard deviation in the readings of 0.02 ml or more. A person of the same experience using correct technique takes about 6 sec and will have a standard deviation in the readings of less than 0.01 ml.

Volumetric Flasks

Volumetric flasks are designed to contain a specified volume of liquid. Since accuracy in their use is not highly dependent on technique, calibration often is unnecessary for work at the level of a part per thousand. The principal source of error is variation in temperature, which causes enough expansion or contraction of aqueous solutions to give errors on the order of 0.1%/5°C.

Before use a volumetric flask should be cleaned thoroughly. The solution to be diluted or the solid to be dissolved is then transferred to it with the aid of a funnel. Often it is more convenient for a solid to be dissolved in a beaker or a conical flask first, and then the resulting solution transferred quantitatively. In the most accurate work the flask is filled to just below the mark and immersed in a water bath at the calibration temperature before final volume adjustment. Alternatively, the temperature of the solution may be taken and a correction made. In careful work droplets of solvent above the meniscus may be removed with a lintless towel after the meniscus has been adjusted but before mixing. Finally, the solution is mixed thoroughly by inverting, shaking, and turning upright at least 10 times.

Pipets

Graduated (Mohr) *pipets* are used for delivering small volumes of liquid with an accuracy of about 1%.

Volumetric pipets are used to deliver precisely a single, definite volume of liquid. The tip must meet stringent requirements,[1] because drainage time is controlled by the diameter of the tip. The amount of liquid delivered depends on how a pipet is used; accuracies of 1 part per 1000 can be attained readily, provided the pipets are used in a reproducible manner.

The proper use of a pipet will be demonstrated. Keep in mind the following points. First clean the pipet so that water drains smoothly from the interior surface. Rinse the interior by drawing a portion of the liquid to be pipetted into the pipet with the aid of a suction bulb, and then tilt and turn the pipet until all the inner surface has been wetted. Discard this portion of solution and repeat the operation twice. Then draw solution above the mark, wipe the tip and stem of the pipet carefully to remove external droplets, and allow the solution to drain until the bottom of the meniscus is even with the calibration mark. Touch the tip momentarily against a beaker wall to remove excess solution. Move the pipet to the receiving vessel and allow the solution to drain freely. During drainage the pipet should be held vertically, with the tip in contact with the container as illustrated in Figure 1-12. Keep the tip in contact for about 5 sec after free solution flow has stopped and then remove it. The remaining liquid in the tip is left there; do not blow out this portion.

---

[1] According to *Circular 602* of the National Bureau of Standards, Washington, D.C.: "The tip should be made with a gradual taper of from 2 to 3 cm, the taper at the extreme end being slight. A sudden contraction at the orifice is not permitted, and the end of the tip must be ground perpendicular to the axis of the tube. The outside edge should be beveled slightly . . . ."

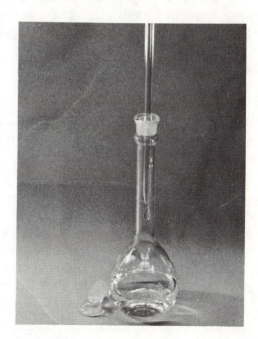

FIGURE 1–12. Position of the pipet during delivery of sample.

FIGURE 1–13. Correct technique for holding a pipet. The index finger, not the thumb, is used to cover the end.

Figure 1-13 shows the proper way to hold a pipet during use. If difficulty is experienced at first with a leaky index finger, try moistening it slightly; too much moisture, however, makes fine control of the outflow rate difficult. After a little practice pipetting will become both rapid and accurate.

A rubber suction bulb is recommended for all pipetting. Particularly, do not pipet poisonous or corrosive solutions by mouth. Rinse the pipet thoroughly after use. Avoid drawing liquid into the bulb. If the bulb is contaminated accidentally, thoroughly rinse and dry it before reuse.

Pipets are marked to deliver a known volume of water at a defined temperature, usually 20 or 25°C, when used in a specified manner. A manufacturer's markings are seldom in error by more than a part or two per thousand when a pipet is used correctly. Since the volume of liquid delivered depends on how the pipet is used, the technique needs to be carefully specified. The calibration exercise outlined below is recommended to compensate for individual variations in handling.

Burets

Burets are designed to accurately deliver measurable volumes of liquid, particularly for titrations. A 50-ml buret, the most common size, has 0.1-ml graduations along its length and can be read by interpolation to the nearest 0.01 ml. Parallax errors in reading are minimized by extension of every tenth graduation around the tube. For reproducible delivery, proper design of the tip is important. Changes affecting the orifice will affect reproducibility; a buret with a chipped or fire-polished tip should not be employed for accurate work. Drainage errors are usually minimized if the tip is constricted so that the meniscus falls at a rate not exceeding ‹0.5 cm/sec.

The accuracy of graduation marks on volumetric burets depends on the diameter of the bore being uniform. The most convenient way to determine actual volumes is to weigh the amount of water delivered. Burets can be purchased already calibrated and accompanied by a calibration certificate. Nevertheless, the calibration operation is best done by the user; it is simple and is an excellent means of learning to use burets correctly.

Because graduations on a 50-ml buret are 0.1 ml apart, volumes between the marks must be estimated. In this estimation the width of the lines must be taken into account. The thickness of a line on a 50-ml buret is usually equivalent to about 0.02 ml and so takes up one fifth of the distance from one mark to the next, as illustrated in Figure 1-14. Generally it is preferable to read the bottom of a meniscus and to take as the value for a given line the point where the meniscus bottom just touches the top of the line. For most people this "edge-to-edge" technique gives consistent results. Figure 1-14 shows readings of various meniscus positions when this system is used. Parallax, another source of error in reading a buret, occurs if the eye is above or below the level of the meniscus. This error can be minimized by use of the encircling markings on the buret as guides to keep the eye level with the meniscus.

The apparent position of the meniscus is significantly affected by the way it is illuminated. Lighting errors are minimized by use of a reading card consisting of a dull-black strip of paper attached to a white background. As shown in Figure 1-15, the card is placed behind and against the buret so that the top of the black portion is flush with, or no more than a scale division below, the bottom edge of the meniscus. For some solutions, such as permanganate and iodine, the bottom of the meniscus may be difficult to see; in such cases the top may be read.

Before using a buret, clean it thoroughly with soap and a buret brush, taking care not to scratch the interior surface. Rinse it well with distilled water; the buret is clean when water drains from the inside surface uniformly without the formation of droplets. Store a buret filled with distilled water and with the Teflon stopcock nut loosened slightly. Before use rinse the buret at least twice with titrant solution and tighten the

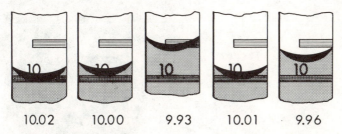

| 10.02 | 10.00 | 9.93 | 10.01 | 9.96 |

FIGURE 1–14. Enlarged sections of a 50-ml buret showing the meniscus at several positions; correct readings are given below each section. (The vertical scale is exaggerated over the horizontal for clarity.)

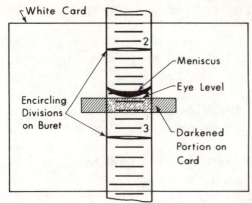

White Card

2

Meniscus

Eye Level

Encircling Divisions on Buret

Darkened Portion on Card

3

FIGURE 1–15. Reading card in correct position.

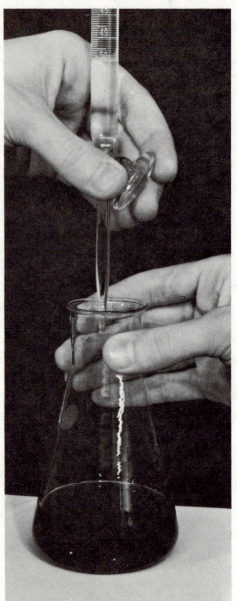

FIGURE 1–16. How to hold a stopcock during use.

Teflon stopcock nut only enough to prevent leakage of solution. After filling with titrant, make certain no air bubbles are present in the tip. Bring the solution level to or slightly below the zero mark, remove the drop adhering to the tip, and wait a few seconds for solution above the meniscus to drain before taking the initial reading.

To provide effective control of the delivery rate when titrating, operate the stopcock with the left hand around the barrel (if right-handed) as in Figure 1-16. The other hand is free to swirl the titration flask.

Graduated Cylinders

Graduated cylinders, or graduates, are used for only approximate measurements. Numerous sizes are available, and by use of the smallest size that will hold the volume being measured an accuracy of about 1% can be attained. A graduated cylinder is not suitable for measurements of part-per-thousand accuracy.

Calibration of Volumetric Equipment

Calibration of volumetric equipment is carried out by weighing the amount of water contained or delivered. Since the density, or weight per unit volume, of water changes about 0.03%/°C at 25°C, the temperature of the water used in the calibration must be known. The actual volume delivered can be obtained from a table relating temperature to volume, such as Table 1-1. Because the density of water is relatively low, a correction must be made for buoyancy. The correction takes into account the upward force exerted on the water and flask by the air they displace. This force corresponds to the net weight of the air displaced by the water, flask, and weights. The density of dry air being 0.0011 g/ml and that of stainless-steel weights 7.8 g/ml, the net correction amounts to about 0.0010 g/ml of water weighed.

TABLE 1–1. VOLUME OCCUPIED BY ONE GRAM OF WATER AT VARIOUS TEMPERATURES[a]

Temp, °C	Vol, ml	Temp, °C	Vol, ml
15	1.0009	26	1.0032
20	1.0018	27	1.0035
22	1.0022	28	1.0038
24	1.0027	29	1.0040
25	1.0029	30	1.0044

[a] If a glass vessel is used at a temperature other than that at which it is calibrated, a small correction factor is required to take into account the expansion of glass. With borosilicate glass this factor amounts to only about 1 part in 10,000 for a 5°C change.

Although errors in placement of buret and pipet markings by the manufacturer are usually small, they cannot be neglected in careful work. For example, if a 10-ml pipet actually delivers 9.990 ml, an error of 1 part per 1000 will be introduced unless the true value is employed.

Procedure

Pipet Calibration

Ask the laboratory instructor for any supplementary instructions, and then clean (with pipe cleaners if necessary) and calibrate the pipets provided. Following the details of pipet use given earlier in this section, weigh the water delivered by each of the pipets. Use a 50-ml conical flask, and perform all weighings on a top-loading balance to the nearest milligram. Calculate the true volume as in the example that follows. Duplicate calibrations should agree to within 0.005 ml. Use the average value of the duplicate calibrations for all subsequent measurements with the pipets.

Example 1–1

Temperature of water	26°C
Weight of flask + water	24.678 g
Weight of flask	14.713 g
Apparent weight of water delivered	9.965 g
True weight of water delivered (see a)	9.975 g
True volume of pipet (see b)	10.007 ml

a. The weight of air displaced by the water is 9.965 ml × 0.0011 = 0.011 g (with the approximation that the density of water is unity). The weight of air displaced by the stainless-steel weights used to weigh the water is (9.965/7.8)0.0011 = 0.0014 g. The net buoyant effect of air on the water weighed is 0.011 − 0.0014 = 0.010 g. The true weight of water delivered is 9.965 + 0.010 = 9.975 g.
b. At 26°C, 1 g of water occupies 1.0032 ml (Table 1-1). The true volume of 9.975 g of water at 26°C is 9.975 × 1.0032 = 10.007 ml.

Buret Calibration

Ask the laboratory instructor for any supplementary instructions. Because water evaporates at an appreciable rate at room temperature (recall the evaporation study in the balance exercise), complete the calibration before starting the calculations.

Clean the buret thoroughly with soap, distilled water, and a buret brush. Mount it in a buret clamp near an analytical balance, or a top-loading balance weighing to the nearest milligram, for calibration. Fill the buret with water that is at room temperature and record the temperature. Withdraw water until the level is at or just below the zero mark. After allowing a few seconds for drainage, take a reading to the nearest 0.01 ml. Weigh a stoppered conical flask to the nearest milligram, and then withdraw water from the buret into it at a rate not exceeding 0.5 ml/sec until the meniscus is within 0.1 ml of the 10-ml mark. Remove the last drop from the tip by touching it to the inside wall of the flask. Stopper the flask and again weigh it to the nearest milligram. Record the buret reading, withdraw another 10-ml portion, and reweigh the flask and contents. Continue until 50 ml is reached. Empty the water from the flask (drying is unnecessary), and repeat the calibration.

A sample set of data is shown in Table 1-2. The first column gives the approximate interval measured, the second the actual readings, and the third the difference between consecutive readings. The fourth column is the weight of the flask plus water, the fifth the difference between consecutive weights, and the sixth the apparent weight as recorded from the balance and corrected for air buoyancy. The seventh column gives the true volume. The eighth column is the true volume minus the apparent volume, and the last column is the cumulative correction from the initial reading near zero.

Check the calculations for the buret calibration by substituting the corresponding values from your table for the letters in the equation

$$A - B + C = D - E + 5(G - F) \qquad (1-3)$$

TABLE 1–2. TABULATION OF DATA TAKEN FOR CALIBRATION OF 50-ML BURET[a]

Approx. Interval	Buret Read-ings	Apparent Volume	Weight	Apparent Weight	True Weight	True Volume	Correc-tion	Cumu-lative Correc-tion
Initial	0.03B		36.450E					
0-10	10.04	10.01	46.420	9.970	9.980	10.01	0.00	0.00
10-20	20.01	9.97	56.381	9.961	9.971	10.00	0.03	0.03
20-30	30.01	10.00	66.362	9.981	9.991	10.02	0.02	0.05
30-40	39.98	9.97	76.264	9.902	9.912	9.94	−0.03	0.02
40-50	49.99A	10.01	86.205D	9.941F	9.951	9.98G	−0.03	−0.01C

[a] Temperature 25°C. The letters after some entries refer to the arithmetic check of Equation (1-3).

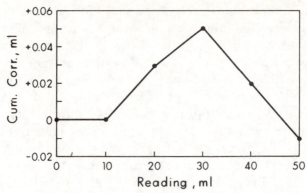

FIGURE 1–17. Calibration chart for a buret.

If the arithmetic is correct, the two sides of the equation should agree within 0.01 ml. After completing this check, ask the laboratory instructor to initial your notebook beside the table.

Duplicate calibration corrections should agree within 0.02 ml. Duplicate cumulative corrections should agree within 0.04 ml at all volumes. If either of these conditions is not met, the technique for reading the buret probably needs improvement, and an additional calibration should be performed. Plot the average cumulative corrections against buret volume in a graph as in Figure 1-17, and attach it to the inside front cover of the laboratory notebook. All subsequent buret readings should be corrected by use of this graph. The net correction for a reading is equal to the correction for the final buret reading minus the correction for the initial reading. This net correction is then added to the net reading. For instance, if the final reading is 40.81 ml and the initial reading 1.03 ml, the net correction interpolated from Figure 1-17 is +0.02 − 0.00, or +0.02 ml. The correct volume is therefore 40.81 − 1.03 + 0.02 = 39.80 ml.

Volumetric-Flask Calibration

Weigh a clean, dry, stoppered flask to the nearest milligram on an analytical balance or a top-loading balance. (The capacity of the balance determines the maximum size of flask that can be calibrated.) Insert a clean, dry funnel into the flask so that the stem extends below the calibration mark of the flask. Fill the flask with water at room temperature to just below the mark. Carefully remove the funnel and add water with a medicine dropper until the bottom of the meniscus coincides with the calibration line. Remove droplets of water present above the line with a lintless towel or a strip of filter paper. Stopper the flask and reweigh it. Immediately after weighing, measure and record the temperature of the water in the flask.

Calculate the true volume of the flask in the same way as for the transfer pipet. Obtain a duplicate calibration by emptying the flask and drying it. Air drying is best if time permits; drying can be hastened by drawing air through the flask with the aid of an aspirator and a piece of glass tubing. Alternatively, *filtered* (clean, grease-free) compressed air can be used. Duplicate measurements should agree to about 5 parts in 10,000.

PROBLEMS

1–8.   What is the correct volume of a 25-ml pipet that delivers 24.912 g of water (apparent weight) at 26°C?

1–9.   In the calibration of a 50-ml pipet by weighing the water delivered, how accurately should the weighings be made if the calibration error from weighing is not to exceed 1 part per 1000?

1–10.   A 10-ml pipet is used without calibration to measure a sample for analysis. If the true pipet volume is 10.046 ml, are the results obtained high or low? By how much?

1–11.   If a pipet calibrated in a normal way were used to deliver liquid while held at an angle of 30° from vertical, would the amount of liquid delivered be more or less than the calibration volume? Why?

1–12.   If the inside diameter of a 10-ml pipet stem is 3.5 mm and the error in adjusting the meniscus is 0.2 mm, what is the corresponding error in the volume delivered? If the overall delivery error is 0.005 ml, what is the most likely source of error?

1–13.   A sample requires addition of the equivalent of 0.1 ml of 3 M H_2SO_4 with an error not exceeding about 3%. How can this be accomplished with a graduated cylinder?

1–14.†   Exactly 0.1000 mole of potassium chloride is dissolved in water and the resulting solution diluted to the mark at 22°C in a flask having a volume of 499.86 ml. What would be the molarity of the solution if the temperature of the solution were changed to: 20°C; 25°C; 30°C? What is the maximum allowable difference between the temperatures of preparation and use if the error is to be kept below 1 part per 1000?

Taking an Aliquot

Whenever a buret or pipet is used to deliver a measured volume of solution, the liquid in that item of glassware prior to measurement should have the same composition as the solution to be dispensed. The following operations are designed to illustrate the minimum effort needed to ensure this.

Fill a pipet with a 5% solution of potassium permanganate and let it drain. Draw some distilled water from a 50-ml beaker into the pipet, rinse, and discard the rinse solution. Determine the minimum *number* of such rinsings required to completely remove the permanganate color from the pipet. If the technique is efficient, three rinsings will suffice. Again fill the pipet with permanganate and proceed as before. This time determine the minimum *volume* of rinse water required to remove the color by collecting the rinsings in a graduated cylinder (less than 5 ml is enough with efficient technique). In the rinsing operations was the water in the 50-ml beaker contaminated with permanganate? If a pink color shows that it was, repeat the exercise with more care.

As a check on your technique, ask the laboratory instructor to observe and comment on the following operation. Rinse a 10-ml pipet several times with a solution of 6 *M* acetic acid. Pipet 10 ml of the acetic acid into a 250-ml volumetric flask. Dilute the solution to volume and mix it thoroughly. Rinse the pipet with the solution in the volumetric flask. Pipet a 10-ml aliquot of the solution into a conical flask.

PROBLEMS

1–15. A sample in a 500-ml volumetric flask is accidentally filled above the graduation mark. A second sample is not available. How can an analysis precise to within 1 part per 1000 still be obtained?

1–16. What is the effect on the amount of solute delivered if, when an aliquot is taken as above: (a) the freshly washed pipet is not rinsed with the concentrated solution; (b) the solution is not mixed adequately after dilution; (c) the same pipet is used throughout, but not adequately rinsed with the diluted solution before the second pipetting step?

1–17. A 0.01136-mole sample was dissolved in water and the resulting solution diluted to the mark in a 100-ml volumetric flask. A series of 10-ml aliquots was pipetted into flasks. How many moles of sample were present in each of the flasks if the calibration volumes of the volumetric flask and pipet were: (a) 100.0 and 10.02 ml; (b) 100.2 and 10.00 ml; (c) 100.2 and 10.02 ml?

1–6   PLANNING LABORATORY WORK

Before starting an experiment, read carefully and understand thoroughly the background and procedure. Knowing each step of the procedure before starting work and why it is included will save time and reduce mistakes. Even as minor an item as dilution of a solution before titrating may be important. If questions arise during study of the experiment, ask the laboratory instructor for advice.

Come to each laboratory period with a work plan. Record sample identification in the laboratory notebook immediately after each sample is obtained. The data page in your notebook should be prepared before starting each experiment. Schedule work to allow for lengthy procedures such as sample drying or dissolution. For example, Experiment 3-2 requires dry sodium carbonate for use in the standardization of a hydrochloric acid solution. Drying can be completed ahead of time while other work is proceeding, and the dry material stored in a tightly closed, labeled container until needed. With effective planning, two or even three experiments may be underway simultaneously.

A minimum of three valid results usually constitutes the basis for selection of a value to report. A sound practice, therefore, is to carry four samples through a determination whenever time and equipment permit. Four results give sufficiently improved statistical validity over three (Table 2-1) that the extra work required is generally worthwhile. A fourth sample also allows for accidental loss of a replicate without jeopardizing the validity of the final result.

A list of the solutions normally provided for each experiment is given in Appendix B. The other required solutions will need to be prepared in appropriate quantities.

It is possible to be overly cautious in making measurements. Time should not be wasted in using a buret when a graduated cylinder is precise enough, or a graduated cylinder when a rough estimation of volume in a beaker is adequate. In measuring solids, do not use an analytical balance when a triple-beam balance is adequate or a triple-beam balance when visual estimation is enough. The reverse, of course, also is true. Never use a graduated cylinder or a triple-beam balance for an operation requiring the precision of a pipet or an analytical balance. Save your patience for situations where accurate measurements are essential, and then be prepared to use care.

1–7 SAMPLING

The purpose of sampling is to obtain a portion of material for analysis that is representative of the average composition of a body of material. This step is occasionally simple, as in the analysis of a solution in a tank for a soluble component, but more often it is a major problem. For example, alloys are frequently nonuniform in composition, and drillings or chips must be taken in such a manner that a cross section is obtained. Solids such as ores, coal, or manufactured chemicals each present special problems, and the number and size of initial samples depend on factors such as the degree of heterogeneity of the bulk material, the particle size, and the accuracy required in the analysis. Reduction of the initially large sample that is then required involves a series of repeated crushing, mixing, and subdividing operations. Because these steps are time consuming and expensive, a final sample is sought with an uncertainty in composition that together with the uncertainty in analysis does not exceed the accuracy needed.

Two major types of sampling errors are encountered. One is related to concentration of the component of interest in various parts of the heterogeneous body; for example, inorganic minerals (ash) tend to be more abundant in coal dust than in larger particles. This error, being of a determinate, or systematic, type, will bias the result to an extent depending on the relative proportion of dust to larger pieces. The other type of error is caused by unsystematic variation in composition of the sample relative to the exact composition of the bulk material. Because these errors are random and can be treated statistically, their effect can be minimized by use of larger samples in the analysis or by an increase in the number of analyses.

The problem of sampling has received much study and is an important part of the total analytical picture. In this book, however, emphasis is placed on the measurement and separation steps; these must be mastered before variations in sample composition can be considered.

1–8 DRYING AND STORING MATERIALS

Some materials pick up varying amounts of water from the atmosphere and so need to be dried before samples are weighed for analysis. Such treatment is especially necessary for finely powdered chemicals that have large surface areas. Metal powders and nonhygroscopic materials do not require drying and can be weighed as received. Drying is usually carried out in an oven at atmospheric pressure, although some substances that tend to decompose on heating can be effectively dried at lower temperatures under reduced pressure.

Once dried, samples must be kept dry until weighing is completed. Uptake of atmospheric moisture or carbon dioxide is reduced if samples are placed in a well stoppered bottle. A weighing bottle is a container with a ground-glass stopper, designed to hold samples during drying, storage, and weighing, though any well stoppered container that is convenient to use on an analytical balance is satisfactory. For added protection, dry samples often are kept in a larger container called a desiccator. Examples of a weighing bottle and a desiccator are shown in Figure 1-18. Desiccators, generally of glass or aluminum, are designed to contain a drying agent such as magnesium perchlorate, phosphorus(V) oxide, calcium sulfate, or calcium chloride. A desiccator maintains a dry atmosphere, provided that the lid is not removed frequently and the desiccant is fresh. Normally it is not used to dry materials, but is convenient for storage and transport in the laboratory of such objects as dried samples or ignited crucibles. When the relative humidity in a laboratory is low, a desiccator can be dispensed with for most samples.

The effectiveness of various drying agents for use in desiccators has been studied. The best is anhydrous magnesium perchlorate, which at

FIGURE 1—18.   An example of a weighing bottle (left) and a desiccator (right).

equilibrium leaves less than a microgram of water per liter of air; phosphorus(V) oxide also is highly effective, but tends to form a surface film of phosphoric acid that limits its rate of water uptake.

SELECTED REFERENCES

W. J. Blaedel and V. W. Meloche, *Elementary Quantitative Analysis*, 2nd ed., Harper & Row, New York, 1963.

D. G. Peters, J. Hayes and G. M. Hieftje, *Chemical Separations and Measurements*, W. B. Saunders Company, Philadelphia, 1974. Chapter 2 provides a discussion of sampling.

F. Trusell and H. Diehl, *Anal. Chem.* **35,** 674(1963). A study of chemical drying agents.

EVALUATION, CALCULATION, AND REPORTING OF EXPERIMENTAL DATA

2–1 HANDLING AND EVALUATION OF EXPERIMENTAL DATA

Statistical thinking will one day be as necessary for efficient citizenship as the ability to read and write.

H. G. Wells

Significant Figures

To indicate the uncertainty in a quantity, the correct number of significant figures must be selected. For instance, if a graduated cylinder is used to measure a certain volume of water, the reported result might be 41 ml—two significant figures. If a buret is used, the result might be 42.21 ml—four significant figures. If the water is weighed on a top-loading balance, the result might be 41.206 ml—here five significant figures are appropriate. In each case no fewer digits should normally be used and no more are justified.

Confusion may arise in the selection of significant figures where zeros are involved. To avoid ambiguity, remember that any zero needed to locate the decimal point is not a significant figure. Thus, 0.005 has one significant figure, but 2.0400 has five. If confusion is likely, numbers should be written in exponential form. For instance, 6.00×10^{10} is unambiguous; it has three significant figures. The number 40,000 has but one significant figure because the four zeros are necessary to locate the decimal point that is understood to be present after the last zero; to prevent ambiguity, it can be written 4×10^4.

When rounding off numbers, if the quantity being rounded off is less than 5, drop it; if it is greater than 5, increase the last significant figure by 1; if it is 5, round to the nearest even number.[1] For example, 4.25, 6.35, and 1.05 are rounded off to 4.2, 6.4, and 1.0. During calculations at least one digit more than the allowed significant figure should be carried along for rounding-off purposes. If possible, avoid rounding off in calculations until the final answer is obtained. With electronic calculators this practice requires little or no additional effort. In analytical work the uncertainty in the last digit of a result may be greater than 1; this is sometimes indicated by lowering the digit a half space. Thus 25.8_5 means that the uncertainty in the 5 is more than 1. In this instance rounding off to 25.8 implies an uncertainty of four parts per thousand, which may be greater than is warranted. See the periodic table and footnote inside the back cover for examples.

Use of Calculators

Electronic calculating machines are quiet and rapid and have largely replaced mechanical models in most laboratories. The so-called four-function models, which add, subtract, multiply, and divide are becoming compact and inexpensive; through their use calculations can be completed more quickly, with fewer errors. Experience is necessary, however, before any calculator can be operated efficiently. To obtain some experience, calculate the value of each of the following, using the appropriate instructions for the calculator available. Report the correct number of significant figures in each answer.

a. 8 × 6/(3 × 4)
b. 3.260 × 8.264 × 0.001264/(0.1376 × 2.165) (Answer: 0.1143)
c. 40.32 × 3.636 × 0.2942/(7.319 × 17.94 × 14.44)
    (Answer: 0.02275)
    40.32 × 3.636 × 0.2942/(7.319 × 17.94 × 14.52)
    40.32 × 3.636 × 0.2942/(7.319 × 17.94 × 14.47)
    40.32 × 3.636 × 0.2942/(7.319 × 17.94 × 14.41)
    40.32 × 3.636 × 0.2942/(7.319 × 17.94 × 14.93)
d. 0.01234/88888 (Answer: 1.388×10^{-7})
e. 1234567 × 9999999 (Answer: 1.234567×10^{13})
f. 123456.789/999.999999 (Answer: 123.456789)

---

[1] Rounding to the nearest even number ensures that results of a series of calculations will not be biased up (if all fives are rounded up) or down (if all fives are dropped).

Some electronic calculators balk at calculations involving many digits, particularly on division.

Errors in Chemical Measurements

In quantitative chemical measurements it is essential to reduce to acceptable levels the errors to which all measurements are subject. Sources of error must be first identified and then the effort required to reduce them to an acceptable level assessed. Usually limitations of time, expense, and technique keep a method from being so accurate and precise that a single measurement can give an answer corresponding to the true value with a high degree of confidence. For this reason, a measurement normally is made in replicate, and the results are treated statistically to provide an estimate of the error of the method. Evaluation of experimental results therefore involves two facets. The first is an assessment of the extent and direction of errors in individual measurements; the second is a statistical treatment of the resulting set of accumulated measurements to give an estimate of the best answer and the degree of confidence warranted in it. Some definitions of terms in common use follow.

The *error* of a measurement is the difference between the value of that measurement and the true value:

$$\text{error} = \text{observed value} - \text{true value} \qquad (2\text{--}1)$$

A measurement is termed *accurate* when its error is small.

Errors may be classified as indeterminate (random) or determinate. *Indeterminate* errors vary in a nonreproducible way around the true value. For a set of measurements they always show two characteristics: large errors occur less frequently than small ones, and the errors are distributed evenly above and below the average value. Random errors can be treated statistically by use of the laws of probability. *Determinate errors* tend to be reproducible from analysis to analysis and generally bias a result in one direction. In general, statistical concepts should not be applied to sets of data that contain determinate errors.

Most errors are partly random and partly determinate. Generally a search for and reduction of determinate errors in analysis is worthwhile. Random errors are not so readily discovered and eliminated, so they are more often treated statistically and taken into account in reporting an analytical result.

The *precision* of a group of measurements refers to how closely individual measurements agree. When the values of a set of measurements are close to each other, they are said to have high precision, and when scattered, low precision. High precision does not necessarily mean high accuracy because a constant error may be present that causes bias in all the results. Low precision indicates errors large enough to produce untrustworthy results unless the central value of a large number of values is taken.

The central value of a set of chemical measurements is most often expressed quantitatively by either the average or the median. The *average,* or *mean,* $\bar{X}$ is defined by

$$\bar{X} = (X_1 + X_2 + \ldots + X_n)/n \qquad (2\text{--}2)$$

where X_1, X_2 ... are the values for individual measurements and n is the total number of measurements. The average is generally the most valid statistical estimate of the true value when determinate errors are absent.

Example 2–1.

The values for a set of measurements of the specific gravity of a solution are 1.012, 1.007, 1.026, 1.020, and 1.014. The average is

$$(1.012 + 1.007 + 1.026 + 1.020 + 1.014)/5 = 1.016$$

To obtain the *median,* arrange the values of a set of measurements in numerical sequence from the largest to the smallest. The median is then the middle value in the set if the number of measurements is odd and the average of the central pair if the number of measurements is even. The median is especially useful for small numbers of measurements having considerable scatter. It has the advantage of being quickly determinable and is less sensitive than the average to deviant values in a set. In the above example the median is 1.014.

The scatter in a set of measurements may be estimated by the *average deviation,* the average of the absolute individual deviations from the mean. For the above set the average deviation is 0.006. When the median is used, the scatter is best expressed by the *probable deviation,* the median of the absolute individual deviations from the median for a set. For the above data the probable deviation also is 0.006.

A more statistically useful measure of the scatter, particularly for random errors, is the *standard deviation s*:

$$s = \sqrt{(d_1{}^2 + d_2{}^2 + d_3{}^2 + \ldots + d_n{}^2)/(n-1)} \qquad (2\text{--}3)$$

where $d_1 = X_1 - \bar{X}$, $d_2 = X_2 - \bar{X}$, and so forth. The standard deviation (or its square, called the *variance*) is used widely in mathematical treatments of data. If the scatter is known to be caused by random error only, calculations based on the standard deviation can give a measure of the degree of confidence justifiable in a set of results.

Errors may be reported in either absolute or relative terms. An *absolute error*, defined by Equation (2-1), has the dimensions of the measurement. The absolute magnitude of an error is not useful, however, unless the observed value also is given. For example, an error of 0.2 mg would not be significant if 10 g of a salt were being weighed for a synthesis, but would be if a 10-mg organic sample were being weighed for microanalytical carbon and hydrogen determinations. To relate the absolute error to the size of the measured quantity, a relative value often is used. *Relative error* is defined as the absolute error divided by the average:

$$\text{relative error} = \text{absolute error/average} \tag{2–4}$$

Similarly, *relative standard deviation* is given by standard deviation/average, $s/\overline{X}$.

Relative quantities have no units and normally are expressed in parts per hundred (pph), parts per thousand (ppt), parts per million (ppm), and so forth—whichever is most suitable. They can be expressed in terms of percentage, but because absolute values for analyses are often given in that form, the "parts per" nomenclature is recommended to avoid confusion.

Example 2–2.

Calculation of deviations for the measurements of specific gravity in Example 2-1.

Value, X	Deviation from Average, $d = |\overline{X} - X|$	Deviation^2 d^2
1.012	0.004	0.000016
1.007	0.009	0.000081
1.026	0.010	0.000100
1.020	0.004	0.000016
1.014	0.002	0.000004
	0.029	0.000217

average deviation = 0.029/5 = 0.006

relative average deviation = 0.006/1.016 = 0.006, or 6 ppt

standard deviation $= \sqrt{0.000217/(5-1)} = \sqrt{0.000054} = 0.007$

relative standard deviation $= 0.007/1.016 = 0.007$, or 7 ppt

In addition or subtraction the absolute error of a result is as large as that of the number with the largest absolute error. For example, if 0.34 is subtracted from 44.6436, the result will have only four significant figures and should be reported as 44.30, not 44.3036. In multiplication and division the relative uncertainty of the result should be of the same order as that of the least accurately known factor. Thus, if 121.463 is multiplied by 0.026, the result should be reported as 3.2, not 3 or 3.16. A convenient guide is to have the uncertainty of the answer lie somewhere between 0.3 and 3 times the uncertainty of the least accurately known number.

Gaussian-Distribution Curve

Everybody believes in the exponential law of errors; the experimenters because they think it can be proved by mathematics; and the mathematicians because they believe it has been established by observation.

E. T. Whittaker and G. Robinson

If a large number of measurements involving only random errors are made, and the frequency with which various values occur is plotted against the values, a symmetrical curve such as shown in Figure 2-1 is obtained. This curve, known as a standard-error, normal-error, normal-distribution, probability, or Gaussian-distribution curve, shows the distri-

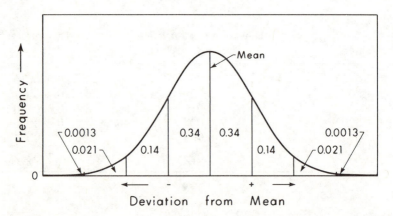

FIGURE 2–1. Gaussian-distribution curve. Vertical lines are shown at 0, 1, 2, and 3 standard deviations from the mean. Numbers refer to fraction of total area in each section.

bution of measurements expected for most analytical systems. The broader the curve, the less precise the measurements. The curve is defined by two parameters: the mean (average), which locates the position of the curve on the horizontal axis, and the standard deviation s, which denotes the spread.[2] For a set of measurements involving only random errors, 68% fall within one, 95% within two, and 99.7% within three standard-deviation units of the average.

Reliability of an Average

Because in real situations the true value is not known, one of the best available approximations to it, the average, is generally used. From the standard deviation of the individual measurements, s, the distribution relation for the Gaussian curve can be applied to estimate the probability that the mean value for an infinite number of measurements (which for unbiased measurements can be called the true value) lies within some range of the measured average. This range is expressed in terms of the confidence limits or confidence interval. For unbiased, independent measurements with a Gaussian distribution, the equation expressing the confidence interval around the measured average within which the true value lies is

$$\text{confidence limits} = \bar{X} \pm ts/\sqrt{n} \qquad (2–5)$$

where t is the tabulated value for a given number of measurements, $n-1$, at some desired probability level. Generally, the convenient values of 90, 95, and 99% probability are used (Table 2-1).

In Table 2-1 the value $n-1$ denotes the degrees of freedom in the system and is used in the calculation of confidence limits of the average for a set of numbers. It can be seen from this table that the size of the correction t in Equation (2-5) depends on the number of measurements used to obtain the standard deviation. Note that, for $n = 10$ or 20, t is already close to the limiting value for $n = \infty$.

Example 2–3.

For the measurements of specific gravity in Example 2-1, $\bar{X}$ is 1.016, s is 0.007, and n is 5. The true value then is 1.016 ±

---

[2] The symbol s is used here to denote standard deviation. When the number of measurements is exceedingly large (ideally n = infinity), the symbol σ is used. The accuracy of s improves with the number of measurements taken. In special cases, such as where a large number of unbiased values are available, the estimate s may be assumed to be σ. Consult the references at the end of Section 2-1 for further information.

TABLE 2–1. t VALUES FOR CALCULATING CONFIDENCE LIMITS
OF AVERAGES

Degrees of Freedom, n-1	t for 90% Probability	t for 95% Probability	t for 99% Probability
1	6.3	12.7	63.7
2	2.9	4.3	9.9
3	2.35	3.2	5.8
4	2.13	2.78	4.6
5	2.02	2.57	4.03
6	1.94	2.45	3.71
7	1.90	2.36	3.50
8	1.86	2.31	3.36
9	1.83	2.26	3.25
10	1.81	2.23	3.17
20	1.72	2.09	2.84
30	1.70	2.04	2.75
∞	1.64	1.96	2.58

$t(0.007/\sqrt{5})$, or $1.016 \pm t(0.0031)$. With 90% confidence the true value falls within the range 1.016 ± 0.007; with 95% confidence, within 1.016 ± 0.009; and with 99% confidence, within 1.016 ± 0.014.

In Example 2-3 we are 90% certain that the true value lies within 1.009 and 1.023, 95% certain that it lies between 1.007 and 1.025, and 99% certain that it lies between 1.002 and 1.030. Clearly, if we want to increase the certainty of the true value lying within some range of the average, the range must be increased. Which confidence level to use depends on the level of certainty needed. The 95% level, which incorporates about two standard-deviation units, is often used in estimates of validity of analytical measurements. Remember that the use of confidence limits is valid only when determinate errors are absent.

Table 2-1 has several applications. In addition to providing confidence limits for a set of measurements, it may be used to test a method of measurement through comparison of an experimentally observed average obtained by that method with the true value. If the difference between $\overline{X}$ and the true value is greater than that normally due to random scatter, a significant determinate (nonrandom) error must be present in the method. If the number of measurements is small, $\overline{X}$ and the true value may be far apart and still not be considered significantly different.

Sometimes the results of different sets of measurements are to be compared. A number of statistical tests are available for this purpose. (Consult the references at the end of Section 2-1.)

Statistics for Small Sets of Numbers

Dealing with the small sets of numbers common to most chemical analyses raises several questions. Foremost is what to select as the best value when one of three or four determinations run on a sample seems to disagree with the others. The first point to consider is the possibility of a determinate error. Objective judgment is required in deciding the extent of such an error and whether rejection is reasonable. When the suspect value cannot justifiably be discarded on the basis of a known determinate error, the difficulty may be caused by unknown determinate error or simply random error. With a small number of observations ($n < 10$), statistical methods applicable to large numbers become less useful or even misleading. Therefore, methods have been developed especially for cases involving few measurements. When only three or four values are available and both variable determinate and indeterminate errors are present, the median is often more reliable than the average in that it is less sensitive to widely divergent values.

Criteria for Rejection of Suspect Values: The Q Test

When the number of values is large, confidence limits can be calculated that enable a suspect value to be judged in relation to its position on the Gaussian-distribution curve. For the purpose of calculating averages the suspect value can then be rejected or retained on the basis of these confidence limits. But, when the number of values is small, the standard deviation becomes inefficient (Table 2-1), and the confidence limits must be spread so widely that almost no values can be rejected.

The best test available for objectively handling data rejection when dealing with small numbers of observations is the Q test:

$$Q_{exp} = \frac{|\text{suspect value} - \text{nearest value}|}{|\text{largest value} - \text{smallest value}|} \tag{2–6}$$

where nearest value refers to the value numerically closest to the suspect value. In calculations of averages and standard deviations the suspect number should be rejected if Q_{exp} exceeds the tabulated value Q_{tab}. Table 2-2 gives Q_{tab} values for small numbers of measurements at the 90% confidence level.

TABLE 2–2. *Q*-TEST VALUES FOR THREE TO TEN MEASUREMENTS AT THE 90% CONFIDENCE LEVEL

n	3	4	5	6	7	8	9	10
$Q_{90\%}$	0.94	0.76	0.64	0.56	0.51	0.47	0.44	0.41

For three values the suspect one usually must be highly divergent before it can be rejected. When results are divergent, additional measurements to provide a more reliable average should be considered. If determinate errors are absent, there is safety in numbers.

Example 2–4.

Four determinations of chloride in a sample give values of 44.28, 44.56, 44.37, and 44.33%. Can the value 44.56% be rejected? For these measurements

$$Q_{exp} = \left|\frac{44.56 - 44.37}{44.56 - 44.28}\right| = \frac{0.19}{0.28} = 0.68$$

Since 0.68 does not exceed the $Q_{90\%}$ value of 0.76 for four measurements, the value 44.56% cannot be discarded. The "best" average to report for this set of results is 44.38%. If the suspect value were 44.87% instead of 44.56%, Q_{exp} would be 0.85. Because the suspect value 44.87% would cause Q_{exp} to exceed $Q_{90\%}$ for four measurements, it would not be included in the average. The "best" average would then be 44.33%.

Criteria for data rejection are not applied to medians, Medians are taken directly from the entire set of data. In Example 2.4 the median would remain 44.35%, whether the suspect value were 44.56 or 44.87%.

PROBLEMS

2–1.   Express the uncertainty of the following measured quantities in parts per thousand: 0.1000 g; 5.12 g; 2.45×10^4 ml; 0.014 cm.

2–2.   Round off to four significant figures: 43.6424; 0.77777; 4.426×10^7; 30000.24.

2–3.   For the following sets of numbers calculate the average deviation and the relative average deviation: (a) 40.00 and 40.13; (b) 0.511, 0.515, and 0.512; (c) 1234, 1247, and 1238.

2–4.   Express the molecular weight of the following compounds to the maximum allowable number of significant figures (using the atomic weights from the periodic table inside the back cover): Ag_2O; B_2O_3; $EuCl_3$; PbS; H_2O.

2–5.   Consider the following set of experimental data: 40.00, 40.15, 40.11, 40.02, 40.07, 40.01, 40.32. What is the median, the probable deviation, the best value for the average, the average deviation, and the standard deviation? What are the 95% confidence limits for the average?

2–6.   Repeat the calculations of Problem 2-5 for the following: (a) 12.34, 12.35, 12.34, 12.37, 12.39, 12.45, 12.37; (b) 0.437, 0.432, 0.429, 0.431; (c) 122.8, 120.2, 121.9, 122.2, 122.6; (d) 87.34, 87.21, 87.46, 87.74.

SELECTED REFERENCES

W. J. Dixon and F. J. Massey, *Introduction to Statistical Analysis,* 3rd ed., McGraw-Hill, New York, 1969. Gives details of advanced statistical tests.

R. J. Flexer and A. S. Flexer, *Programmed Reviews of Mathematics,* Vol. 6, Introduction to Statistics, Harper & Row, New York, 1967. An approximately two-hour programmed introduction to the basic concepts of statistics. Simple and straightforward.

W. S. Gossett, *Biometrika* **6,** 1(1908). This famous paper introduced the *t* test and *t* distribution. It was published under the pseudonym of Student, and is known as Student's *t* distribution. The table of *t* values as we know them today was provided by R. A. Fisher [*Metron.* **5,** 90(1926)].

H. A. Laitinen and W. E. Harris, *Chemical Analysis,* 2nd ed., McGraw-Hill, New York, 1974, Chapter 26.

D. G. Peters, J. Hayes, and G. M. Hieftje, *Chemical Separations and Measurements,* W. B. Saunders Company, Philadelphia, 1974, Chapter 2. Discusses statistical treatment of data.

W. Weaver, *Lady Luck,* Doubleday (Anchor Book), New York, 1963. A delightful introduction to the ideas of statistics.

Analytical Chemistry **44,** 2420(1972). A compilation of recommended terms and symbols for reporting analytical data.

2–2 UNKNOWN SAMPLES AND EVALUATION OF RESULTS OF ANALYSIS

A convenient way to report the results of laboratory analyses is with cards of the type shown in Figure 2-2.

The grade for each experiment depends on how close the reported value is to the presumably correct value. On a 5-point grading scale a grade of 5 normally denotes a result within 90 or 95% confidence limits of the correct value and equivalent in quality to the work of an experienced analyst. Other grades are assigned according to the deviations in increasing geometric progression. For example, in a typical case for an analysis a grade of 4 denotes a result deviating up to 2 times that allowed for a 5, 3 a result deviating up to 4 times, and 2 a result deviating up to 8 times. A grade of 1 is assigned to a result outside the 2 limits. Grades of 4 and 5 on this scale may be considered excellent work, while grades of 1 and 2 indicate defective work.

REPORT OF ANALYSIS Chem. 312

Determination of ....

Sample Code No. .... Analysis ....

Name .... Date ....

Only the result "Analysis" is graded. For completeness report:

Average .... Median ....

Also if suitable: Av. Dev. .... Prob. Dev. ....

Alternate or Add'l Items

Room ....

Locker ....

Det of ....

FIGURE 2–2. A 3- by 5-in. report form for analytical results.

Confidence limits for grading can be calculated from a combination of the standard deviations of all the measurements made in an experiment. For example, the standard deviation of a properly taken buret reading by a student is 0.014 ml. The uncertainty (standard deviation) in the volume of an uncalibrated 50-ml volumetric flask is 0.04 ml and of a calibrated 50-ml volumetric flask 0.02 ml. Where standard deviations are not available for an operation or a piece of equipment, reasonable assumptions can be made, such as that the standard deviation of a weight obtained by difference is 0.14 mg and of a typical end-point selection is 0.03 ml, and that the uncertainty in the correct value of the material being determined is 1 part per 1000. The standard deviations of all operations in a given experiment such as weighing, buret reading, buret calibration, use of a volumetric flask, pipetting, choice of end point, reading a spectrophotometer, measuring a peak area, and sample error are combined statistically to obtain the confidence limits. Thus, in Experiment 3-1, 90% confidence limits of 3.5, 2.3, and 2.0 parts per 1000 can be assigned to samples containing 20, 40, and 60% chloride. The 90% confidence limits for typical samples of substances whose analyses are described in this book are: chloride, 0.1%; carbonate, 0.3%; calcium, 0.2%; iron, 0.15%; iodometric copper, 0.2%; glycol, 0.006 g; sulfur, 0.04%; cerium, 0.1%; and salt, 0.025 meq. From this discussion it can be seen that a grade of 5 on the scale above connotes results of unusual quality and is difficult to attain consistently.

Formulas for the calculation of results are sometimes provided in an experiment. When feasible calculate the results for each experiment independently first, and use formulas only as a check. Calculation of results is a part of every analysis, and calculation errors are usually disastrous. The usefulness of an analysis depends on the accuracy of the value reported.

Follow-Up Reports

When a failing grade (1 or 2 on the scale described above) is obtained, it is important to attempt to locate the error as soon as possible so that similar errors can be avoided in the future. In this matter the instructor can help, although it is best if the student can identify the difficulty for himself. To aid in a systematic investigation of poor results, the student, individually at first, then with the instructor, may complete a follow-up report similar to the one shown in Figure 2-3.

Many failing results are caused by errors in calculation. Common sources of such mistakes include the use of incorrect formulas, use of incorrect molecular weights, misplaced decimal points, omission of factors in calculation, and too few significant figures. Inadequate record keeping is another frequent cause of poor results. A positive correlation exists between the quality of records and the quality of experimental work. If poor results for several experiments are otherwise inexplainable, the work probably is being performed without adequate comprehension of the background. Poor agreement among replicates often serves as a warning that results will be unsatisfactory.[1]

---

[1] A recent survey of follow-up reports at the University of Alberta, based on combined student-instructor assessment of poor results, showed that of the approximately 16% of analyses that earned a grade of 1 (out of 5) 56% had errors in calculation, 12% appeared to involve errors in procedure or technique, and 18% had poor agreement of replicate analyses. In 14% of the cases (2% of the total analyses reported) the cause could not be determined. For grades of 2 out of 5 (about 12% of all the analyses reported) 25% had 1 or more errors in calculation, 20% had errors in procedure or technique, 23% had poor agreement of replicates, and 32% (about 4% of the total analyses reported) had no ascertainable source of error.

FOLLOW UP ON A GRADE OF 1 OR 2

Name .... Det'n of .... Code No. ....

Orig. individual results ....

Orig. result rept'd. .... Orig. grade .... Recalc. result ....
                                     (high) (low)   (if calc. error)
(Possible) (probable) (undoubted) explanation of poor result:

When above section completed ask lab instructor to complete items below:
Lab notebook—Original, systematic, complete: (exc.) (sat.) (fair) (poor)
Comment on whether above explanation is reasonable:

Instructor's Signature ....

FIGURE 2–3.   Example of a follow-up report.

2–3   CALCULATIONS IN VOLUMETRIC ANALYSIS

Solution Concentrations

The most common ways of expressing solution concentration in volumetric analysis are molarity, normality, and titer. All express the amount of a solute present per unit volume of solution. *Molar* concentration is defined as the number of gram molecular weights (moles) of a substance dissolved in 1 liter of solution.[1] A one-molar (1 *M*) solution of silver nitrate contains 1 mole of silver nitrate (169.87 g) per liter of solution, or 1 millimole (mmole) per milliliter of solution. *Normal* concentration is defined as the number of gram equivalent weights (equivalents) of a substance dissolved in 1 liter of solution, or the number of milliequivalents per milliliter. Since the gram equivalent weight of a substance varies with the reaction in which it is involved, a single solution may have several different normalities. Because of this ambiguity, normalities are not used in this book. The *titer* of a solution is the weight of a substance that is equivalent to, or reacts with, a unit volume of the solution. Titer values are used frequently in routine analysis because they simplify calculations. For example, in Experiment 3-1 the chloride titer of the silver nitrate solution is (0.1 *M*) (mol wt Cl/1000), or (0.1) (0.03545). The weight of chloride in a sample can be determined simply by multiplying the milliliters of silver nitrate required to titrate the sample by the value of the chloride titer.

Solute concentrations may be expressed also in terms of weight, that is, as the amount present per unit weight of solution or solvent. Weight units include *molality,* defined as the number of gram molecular weights of a substance dissolved in 1 kg of solvent; *weight per cent,* defined as [wt solute/wt solution] (100); and *weight molarity,* defined as the number of gram molecular weights of a substance per kilogram of solution. Molality and weight per cent are generally not convenient to use in volumetric analysis, but weight molarity is useful when titrations are carried out by measurement of the weight of solutions rather than volume, as in Experiment 7-2.

Trace concentrations are sometimes expressed in units of parts per million (ppm) or parts per billion (ppb) by weight. Thus 1 mg of calcium in 1 liter of water corresponds to a calcium concentration of 1 ppm.

Calculation of Percentages

Volumetric calculations most often involve conversion of experimentally obtained data—generally volumes, concentrations, and weights—to the percentage of a substance present in a sample. A useful relation is

---

[1] The term *formality* is sometimes used in place of molarity to emphasize that the nature of the species present in solution is not specified.

that the number of moles of material in a sample multiplied by the molecular weight w equals the grams g of material present:

$$g = (\text{moles}) (w) \qquad (2\text{--}7)$$

The number of moles of material in a given volume of solution is obtained by multiplying the volume V by the molarity M. Therefore, the weight of material present in a dissolved sample is given by

$$g = (V) (M) (w) \qquad (2\text{--}8)$$

where V is in liters. In titrations, V is more conveniently expressed in milliliters; then

$$g = V_{ml}Mw/1000 \qquad (2\text{--}9)$$

   In stoichiometry calculations, cultivate the habit of leaving the arithmetic until last. When a numerical answer is not required, report the answer in factorial form. Thus the weight of sodium sulfate in 45 ml of 0.3 M solution would be reported as

$$g\ Na_2SO_4 = \frac{45}{1000}\ (0.3)\ (\text{mol wt } Na_2SO_4)$$

The following examples illustrate calculations often encountered in volumetric work.

Example 2–5.

   Preparation of a standard solution. How much silver nitrate is required to prepare 250.0 ml of a 0.1000 M solution?

g $AgNO_3$ = (250.0 ml) (0.1000 mole/liter)

$$(169.9\ g\ AgNO_3/\text{mole})/(1000\ ml/\text{liter})$$

   = 4.247 g

Example 2–6.

   Titration of a chloride sample. The chloride in a 0.5185-g sample required 44.20 ml of 0.1000 M $AgNO_3$ for titration. What is the percentage of chloride in the sample?

$$\%\ Cl = \text{wt Cl } (100)/(\text{wt sample})$$

$$\text{wt Cl} = (\text{moles Cl}) (\text{at. wt Cl})$$

One mole of silver nitrate reacts quantitatively with 1 mole of chloride in the balanced equation, so the number of moles of chloride present in the sample equals the number of moles of silver used in the titration. Since

$$\text{moles AgNO}_3 = (\text{ml AgNO}_3)\,(M\text{ AgNO}_3)/1000$$

the above relations can be combined to give

$$\%\text{ Cl} = \frac{(\text{ml AgNO}_3)\,(M\text{ AgNO}_3)\,(\text{at. wt Cl})}{(\text{wt sample})\,(1000)}\,(100)$$

$$\%\text{ Cl} = \frac{(44.20)\,(0.1000)\,(35.45)\,(100)}{(0.5185)\,(1000)} = 30.22\%$$

If a solution of silver nitrate is to be standardized against a pure salt such as sodium chloride, the same expression can be used to determine the molarity of the silver nitrate by substituting for % Cl the correct value for pure NaCl, 60.66%, and solving for M AgNO$_3$.

Example 2–7.

Calculation of titrant volume needed for a sample. How many milliliters of 0.1000 M AgNO$_3$ are required to react with 0.4250 g of BaCl$_2 \cdot 2$H$_2$O? Since 1 mole of BaCl$_2$ reacts with 2 moles of AgNO$_3$ in the balanced equation,

$$\text{moles AgNO}_3 = (2)\,(\text{moles BaCl}_2 \cdot 2\text{H}_2\text{O})$$

$$= (2)\,\frac{\text{g BaCl}_2 \cdot 2\text{H}_2\text{O}}{\text{mol wt BaCl}_2 \cdot 2\text{H}_2\text{O}}$$

$$\text{ml AgNO}_3 \text{ required} = \frac{(\text{moles AgNO}_3)\,(1000)}{M\text{ AgNO}_3}$$

$$\text{ml AgNO}_3 = \frac{(2)\,(0.4250)\,(1000)}{(244.3)\,(0.1000)} = 34.79$$

SELECTED REFERENCE

H. Diehl, *Quantitative Analysis,* Oakland Street Science Press, Ames, Ia., 1970. Stoichiometry in gravimetric and volumetric systems is treated clearly and simply.

Chapter 3

TITRIMETRIC ANALYSIS: PRECIPITATION, ACID-BASE, AND COMPLEXATION

The road to wisdom?—Well, it's plain and simple to express:

Err
and err
and err again
but less
and less
and less

Piet Hein

Chapters 3 and 4 contain primarily experiments designed to give experience in techniques of volumetric wet-chemical methods. The basic operations of volumetric analysis (introduced in Chapter 2) are applied to acid-base, precipitation, complexation, and oxidation-reduction titrations.

Volumetric methods of analysis permit a large variety of substances to be measured rapidly, conveniently, and precisely and are therefore emphasized in two chapters. The methods require only simple equipment and are applicable to a wide range of sample concentrations. Also, the quantitative volumetric techniques introduced here are needed in most of the instrumental and separation methods encountered in later chapters.

3–1 PRECIPITATION TITRATIONS: ARGENTIMETRIC DETERMINATION OF CHLORIDE

Background

In this experiment the reaction between silver ion and chloride,

$$Ag^+ + Cl^- \rightleftarrows AgCl_{(s)} \qquad (3-1)$$

is used to illustrate volumetric precipitation analysis, that is, a titration in which the reaction product is an insoluble substance. Reaction (3-1) is rapid and the stoichiometry is accurate—so accurate, in fact, that this reaction was used in the early determination of several atomic weights.

In this experiment two different methods of end-point detection are compared. One is the Mohr method, which uses the appearance of silver chromate to mark the end point. The other is the Fajans method, which uses the formation of a colored adsorbed layer on the silver chloride precipitate.

These titrations illustrate two important techniques—the direct preparation of a standard solution and the taking of aliquots. The operations of pipetting, using a buret, and simple quantitative transfer of solutions are emphasized. The taking of aliquots frequently saves time and is especially useful when the sample size is small. A disadvantage in the use of aliquots is that errors in the initial weighing and dilution steps are not revealed by replicate titrations.

Silver nitrate is an excellent primary standard. It may be dried at 110°C for 1 to 2 hr to remove adsorbed surface water, if desired, although drying is normally unnecessary. The solid tends to discolor somewhat upon heating, apparently because of reduction of silver (I) to the metal by traces of organic material. The error is negligible unless the salt has been contaminated by a considerable amount of organic matter. Solutions of silver nitrate are stable, although slow photoreduction occurs if protection from prolonged exposure to bright light is not provided.

The Mohr Method

This analytical method, first proposed by Mohr in 1856, is one of the oldest still in use. It is based on the formation of an orange-red precipitate of silver chromate at the end point. When a mixture of chloride and chromate is titrated with silver nitrate, silver chloride precipitates first, being less soluble than silver chromate. After the chloride in solution has precipitated, the first excess of silver ion precipitates silver chromate to mark the end point.

For the end point and the equivalence point to coincide, the chromate ion concentration in theory should be adjusted so that the solubility product of silver chromate is just exceeded when the silver and chloride ion concentrations become equal. In practice, a chromate concentration of about 0.002 M gives more precise results than the value of about 0.02 M calculated from the solubility products because the yellow color of chromate in solution tends to mask the first appearance of the red precipitate.

The end-point error in the Mohr method is appreciable and depends in part on the amount of silver chloride precipitate present. Increasing amounts of silver chloride make it more difficult to see the first traces of silver chromate. Errors can be minimized by standardizing the silver nitrate against primary-standard sodium chloride. If the amounts of chloride present in the standard and sample titrations are about equal, end-point errors will cancel each other. If silver nitrate is used directly as a standard, the error can be reduced by titration of a blank containing a white insoluble compound, such as calcium carbonate, to simulate the silver chloride precipitate. This is the procedure employed in this experiment.

A *blank* titration is one in which all the reagents are added, but the sample is not. It compensates for errors due to titratable impurities in the reagents and to late end points, but seldom compensates for errors due to early end points. The best way to correct for errors resulting from early end points and some titratable impurities in reagents is to standardize the titrant against a sample of the same material of known purity such as sodium chloride, as mentioned above.

The Fajans Method

The end point in many precipitation titrations also can be detected through use of an adsorption indicator. Adsorption indicator methods were first developed by Fajans. Dichlorofluorescein, a weak organic acid, may be used for the titration of chloride with silver (I). The dichloro-fluoresceinate anion tends to be adsorbed on the surface of precipitated silver chloride, but not so strongly as chloride. In the first stages of a titration, chloride ions in solution are adsorbed preferentially on the surface of the precipitate. As the titration proceeds, the chloride ion concentration decreases until it approaches a small value at the equivalence point. After the equivalence point a slight excess of silver ion is present, and the attraction between adsorbed silver ion and dichloro-fluoresceinate anion causes the latter to be adsorbed on the surface of the precipitate. This adsorbed layer is pink, and the appearance of this pink color on the precipitate is taken to be the end point. Because the color change takes place on the surface of the precipitate, it is sharper if the surface area is large. Certain compounds, such as dextrin, keep precipitates of silver halides from coagulating and provide the needed surface area.

In acidic solutions neither chromate nor dichlorofluoresceinate is a satisfactory indicator because, being anions of weak acids, both form protonated species that will not function as indicators. Therefore buffer is often added to control the pH during the titration.

Preparation for Experimental Work

Before starting laboratory work, read the experimental procedure carefully and set up a summary data page in your laboratory notebook. (Follow this practice for every experiment.)

Procedure (median time 4.5 hr)

Preparation of 0.1 M AgNO$_3$

Weigh accurately on an analytical balance a clean, dry 50-ml beaker. Transfer the beaker to a triple-beam balance and weigh into it enough AgNO$_3$ to prepare 250 ml of 0.1 M solution.[1] Reweigh the beaker plus compound on the analytical balance. Dissolve the AgNO$_3$ in a small portion of chloride-free distilled water.[2] Transfer this solution quantitatively to a clean 250-ml volumetric flask with the aid of a funnel, and dilute to volume with distilled water. Mix well. Store protected from light.

Preparation of Sample Solution

Weigh, to the nearest 0.1 mg in a 50-ml beaker, from 2.0 to 2.1 g of a dry sample and dissolve in a small volume of distilled water. Transfer the solution quantitatively to a 100-ml volumetric flask, dilute to volume, and mix well. Pipet 10-ml aliquots (Section 1-5) of this solution into each of several 200-ml conical (Erlenmeyer) flasks, using a calibrated pipet.

Mohr Titration

To two or three of the aliquots add about 40 ml of distilled water, about 0.5 g of NaHCO$_3$, and 1 ml of 5% K$_2$CrO$_4$ solution.[3] Titrate with AgNO$_3$ solution, swirling the flask continuously to prevent a local excess of Ag$_2$CrO$_4$ from being occluded by the AgCl precipitate. The end point is the first permanent appearance of red-orange in the yellow chromate

---

[1] The weight of AgNO$_3$ taken should be within 5% of the amount required to make 250 ml of 0.1 M solution. AgNO$_3$ is extremely corrosive, so it must not be handled near an analytical balance.

[2] Chloride-free distilled water should be used throughout this experiment. Anionic impurities in water are most readily removed by passage through a column containing an anion-exchange resin in the hydroxide form. If desired, both cationic and anionic impurities can be removed in one step by use of a mixture of cation-exchange resin in the hydrogen form and anion-exchange resin in the hydroxide form.

[3] If the samples are neutral, pH adjustment is unnecessary. Because H$_2$CO$_3$ may be present in the solution through absorption of CO$_2$, NaHCO$_3$ is added to raise the pH above 6.

solution.[4] Determine an indicator blank by repeating the titration, substituting about 2 g of powdered $CaCO_3$ for the aliquot.[5] Subtract the blank volume from the volume of $AgNO_3$ used in each Mohr titration.

Fajans Titration

To the remaining aliquots add about 1 ml of 0.5 *M* acetic acid-0.5 *M* sodium acetate buffer and 0.3 g of dextrin. Titrate with $AgNO_3$ solution to within about 1 ml of the end point, and then add 5 drops of 0.1% dichlorofluorescein solution.[6] Immediately continue the titration until the precipitate becomes pale pink, swirling the flask constantly. No blank correction is necessary.

Calculate the percentage of chloride in the sample and submit a report of the analysis.

Calculations

Because a large number of poor results are caused by errors in calculation, all calculations should be checked. Many hours of careful laboratory work can be thrown away through a simple mistake in arithmetic or an incorrect setup for the calculations. In the early experiments the necessary calculations are outlined in detail. As the course progresses, the amount of detail given decreases until you will be expected to do all calculations on your own with only an occasional suggestion. It is important to understand the reasons behind each step in every calculation. Do not hesitate to ask the instructor to explain any steps that are not clear.

Example 3–1.   (See also Section 2-3.)

$$\% \, Cl = \frac{wt \, Cl}{wt \, sample \, in \, aliquot} (100)$$

$$wt \, Cl = (moles \, Cl) \, (g \, Cl/mole)$$

---

[4] If a gradual indicator change makes end-point selection uncertain, record the buret reading. Then add another drop and note any change. Record again and repeat. Select as the end point the reading corresponding to the drop that causes the greatest change.

[5] The $CaCO_3$ is used to simulate the white AgCl precipitate. If no precipitate is present in the blank titration, the end point appears from 0.02 to 0.10 ml early. Standardization of the $AgNO_3$ solution against standard NaCl is not required here because of the additional check provided by the Fajans titrations; if the Fajans titrations are not carried out, a set of standard NaCl samples should be prepared and run.

[6] Dichlorofluorescein promotes photoreduction of silver in the AgCl precipitate. The resulting finely divided dark silver obscures the end point. Addition of the indicator near the equivalence point minimizes this difficulty. If the approximate location of the end point is unknown, indicator must be added early in the first titration, and the approximate end-point volume determined. Photodecomposition may cause some error even when the indicator is added late in the titration, so direct sunlight on the titration flask should be avoided.

Since 1 mole of Ag^+ reacts with 1 mole of Cl^- in Equation (3-1),

$$\text{wt Cl} = \frac{(\text{ml AgNO}_3)\,(M\,\text{AgNO}_3)\,(35.45\,\text{g/mole})}{1000}$$

$$\%\,\text{Cl} = \frac{(\text{ml AgNO}_3)\,(M\,\text{AgNO}_3)\,(35.45\,\text{g/mole})}{(1000)\,(\text{wt sample in aliquot})}\,(100)$$

The weight of sample in the aliquot taken for each titration is given by

$$(\text{initial wt of sample})\,\frac{\text{calibration volume of 10-ml pipet}}{100\,\text{ml}}$$

Note that the molarity is expressed in moles per liter rather than millimoles per milliliter, and the formula weight of chloride in grams per mole rather than grams per millimole. Accordingly, the volume must be converted to liters by dividing by 1000. This convention is used throughout the book, even though other combinations of units are often used. No one exclusive way must be followed; use the one you find most convenient. A check of the units at the end will frequently reveal errors. A check of units in the first part of this example gives

$$\%\,\text{Cl} = \frac{(\text{ml})\,(\text{moles}/1)\,(\text{g/mole})}{(\text{ml}/1)\,(\text{g})}\,(100)$$

The molarity of the $AgNO_3$ is calculated from the expression

$$M\,\text{AgNO}_3 = \text{moles/liter}$$

$$\text{moles AgNO}_3 = \frac{\text{g AgNO}_3}{\text{g AgNO}_3/\text{mole}} = \frac{\text{wt AgNO}_3}{169.87}$$

$$\text{liters AgNO}_3 = 250.0/1000$$

$$M\,\text{AgNO}_3 = \frac{\text{g AgNO}_3}{(169.87)\,(0.2500)}$$

PROBLEMS

3–1.   A 0.3297-g sample of a mixture of potassium nitrate and sodium chloride required 44.25 ml of 0.0995 M silver nitrate for titration. What is the percentage of chloride present?

3–2.   A calcium chloride brine solution was analyzed by titration with silver nitrate. If 38.92 ml of 0.1020 M titrant was required for a 0.5229-g sample, what is the percentage of chloride in the brine?

3–3.   What volume of 0.1009 M silver nitrate is required to titrate a 0.2212-g sample of pure potassium bromide?

3–4.   A sample of pure lithium iodide required 42.89 ml of 0.1014 M silver nitrate for titration. What was the weight of the sample?

3–5.   A 25.00-ml sample of 0.1000 M sodium chloride solution is titrated with 0.0700 M silver nitrate. (a) Calculate the concentration of chloride ion in solution in molarity and pCl ($-\log[Cl^-]$) after 5, 30, 45, and 50 ml of silver nitrate solution has been added. (b) Calculate [Cl⁻] and pCl at the equivalence point in the titration. (c) Plot pCl against volume of silver nitrate for the calculated points.

3–6.   Calculate the theoretical concentration of chromate ion needed to make the end point and the equivalence point coincide in the titration of a 0.05 M solution of potassium thiocyanate with 0.05 M silver nitrate.

3–7.†   The chloride content of a sample is known to be in the range 30 to 40%. What are the minimum and maximum sample sizes that may be taken if the volume of 0.1000 M silver nitrate used in the titration is to be in the range 35 to 50 ml?

3–8.   A chemist wished to find the volume of a large barrel and had no equipment for measuring the amount of water needed to fill it. He added 404 g of sodium chloride to the barrel and filled it with water. After the solution was mixed, a 100.0-ml sample required 36.66 ml of 0.0505 M silver nitrate for titration. What was the volume of the barrel?

SELECTED REFERENCES

K. Fajans and O. Hassel, *Z. Elektrochem.* **29**, 495(1923). Original paper on the use of fluorescein as adsorption indicator for the titration of silver chloride. Dichlorofluorescein was suggested a few years later by Kolthoff, Lauer, and Sunde as being somewhat better than fluorescein in solutions of pH less than 4 because it is a stronger acid.

I. M. Kolthoff, E. B. Sandell, E. J. Meehan, and S. Bruckenstein, *Textbook of Quantitative Inorganic Analysis,* 4th ed., Macmillan, New York, 1969, pp 719 and 796.

F. Mohr, *Ann.* **97**, 335(1856). Original paper on chromate indicator for chloride determination.

D. G. Peters, J. Hayes, and G. Hieftje, *Chemical Separations and Measurements,* W. B. Saunders Company, Philadelphia, 1974, Chapter 8.

R. W. Ramette, *J. Chem. Educ.* **37**, 348(1960). An outline of the experimental determination of the equilibrium constants for the silver chloride system, including the formation of $AgCl_2^-$ at high ratios of chloride to silver.

T. W. Richards, *Chem. Rev.* **1**, 1(1924). A summary of the determination of atomic weights by chemical methods.

3–2 ACID-BASE TITRATIONS: DETERMINATION OF CARBONATE BY TITRATION WITH HYDROCHLORIC ACID

Background

Carbonate Equilibria

In this experiment a solution of hydrochloric acid is prepared, standardized against pure sodium carbonate, and used to determine the percentage of carbonate in a sample.

An aqueous solution of hydrochloric acid is almost completely dissociated into hydrated protons and chloride ions. Therefore, in a titration with hydrochloric acid the active titrant species is the hydrated proton. This species is often written H_3O^+, although the actual form in solution is more correctly $(H_2O)_nH^+$. For convenience we designate it simply H^+.

Carbonate in aqueous solution acts as a base; that is, it is able to accept a proton to form bicarbonate ion:

$$CO_3^= + H^+ \rightleftarrows HCO_3^- \tag{3-2}$$

Bicarbonate is able to combine with another proton to form carbonic acid:

$$HCO_3^- + H^+ \rightleftarrows H_2CO_3 \tag{3-3}$$

Equilibrium expressions for the dissociation of bicarbonate and carbonic acid may be written

$$K_2 = \frac{[H^+]\,[CO_3^=]}{[HCO_3^-]} \tag{3-4}$$

and

$$K_1 = \frac{[H^+]\,[HCO_3^-]}{[H_2CO_3]} \tag{3-5}$$

where K_1 and K_2 are the first and second acid dissociation constants for H_2CO_3; the experimentally determined values are $K_1 = 3.5 \times 10^{-7}$ and $K_2 = 5 \times 10^{-11}$.

When successive protonation reactions such as (3-2) and (3-3) occur, the extent to which the first reaction proceeds before the second begins depends on the difference between the two acid dissociation constants. By combination of Equations (3-4) and (3-5) with those for charge and mass balance, $[H^+]$ can be calculated for any ratio of hydrochloric acid to initial carbonate concentration, that is, at any point on a titration curve of carbonate with hydrochloric acid. Because complete and rigorous solution is time consuming, here only procedures for calculating the pH at several convenient points in a titration of 0.1 M sodium carbonate with 0.1 M hydrochloric acid (Figure 3-1) are covered briefly. An analytical textbook should be consulted for a more detailed discussion of this topic.

pH at Point A in Figure 3-1. At point A no acid has been added, and only sodium carbonate is present in solution. The pH is determined by the extent of carbonate reaction with water to give HCO_3^- and OH^-:[1]

$$CO_3^= + H_2O \rightleftarrows HCO_3^- + OH^-$$

(3–6)

Here water acts as an acid, providing a proton to carbonate ion, the base. The equilibrium constant for this reaction may be written

$$K_b = \frac{[HCO_3^-] \, [OH^-]}{[CO_3^=]}$$

(3–7)

---

[1] Reactions of ions of a solute with water often are called hydrolysis reactions. They are more properly considered, however, as simply another example of a Bronsted acid-base reaction in which water acts as an acid or a base.

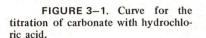

FIGURE 3–1. Curve for the titration of carbonate with hydrochloric acid.

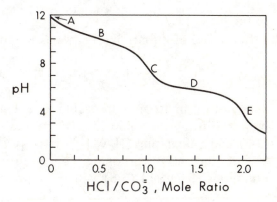

Multiplying the right side of Equation (3-7) by $[H^+]/[H^+]$, we see that K_b is equal to K_w/K_2, where K_w is the dissociation constant for water,

$$K_w = [H^+][OH^-] = 10^{-14} \text{ at } 24°C \tag{3–8}$$

and K_2 is the second dissociation constant for carbonic acid [Equation (3-4)]. If the initial concentration of carbonate and the values of K_w and K_2 are known, $[OH^-]$ can be calculated from

$$\frac{K_w}{K_2} = \frac{[HCO_3^-][OH^-]}{[CO_3^=]} \tag{3–9}$$

Assume that the equilibrium for Equation (3-6) lies far to the left, so that the carbonate ion concentration is still essentially 0.1 M. Since bicarbonate and hydroxide are formed in equimolar amounts,

$$[HCO_3^-] = [OH^-] \tag{3–10}$$

Substitution of numerical values and Equation (3-10) in Equation (3-9) gives

$$\frac{10^{-14}}{5 \times 10^{-11}} = \frac{[OH^-]^2}{0.1} \tag{3–11}$$

and

$$[OH^-] = 4.5 \times 10^{-3} \, M \tag{3–12}$$

From Equation (3-8)

$$[H^+] = \frac{10^{-14}}{4.5 \times 10^{-3}} = 2.2 \times 10^{-12} \, M \tag{3–13}$$

so the pH is 11.7.

In our use of Equation (3-7) we assume that the reaction

$$HCO_3^- + H_2O \rightleftarrows H_2CO_3 + OH^- \tag{3–14}$$

does not occur to an appreciable extent; that it does not can be verified by substituting the value for $[H^+]$ found in Equation (3-13) in Equation (3-5) and calculating $[H_2CO_3]$. If $[H_2CO_3]$ is found to be greater than 5% of the total carbonate concentration, the $[H^+]$ calculated from Equations (3-7) and (3-8) will be appreciably in error. In this case the expression should be solved either exactly, by including all species (which is tedious), or by successive approximations. Calculation shows that $[H_2CO_3]$ at Point A is negligibly small, so our assumption is valid. The additional assumption that $[CO_3^=]$ is essentially 0.1 M also is confirmed because Equations (3-10) and (3-12) show that $[HCO_3^-]$ is less than 5% of $[CO_3^=]$.

Note from this discussion that $\dfrac{K_w}{K_2} = K_b$, or $K_w = K_2 K_b$. Thus, if K_a for an acid HA is known, K_b for the corresponding base A^- can be calculated in aqueous solutions. An acid HA and base A^- are called a *conjugate* acid-base pair; HA is the conjugate acid of A^- and A^- the conjugate base of HA.

pH at Point B. At Point B in Figure 3-1, ½ mole of hydrochloric acid has been added for each mole of carbonate. The solution now contains an equimolar mixture of carbonate and bicarbonate. We can calculate the pH at this point by rearranging Equation (3-4) to

$$[H^+] = \frac{[HCO_3^-]\, K_2}{[CO_3^=]} \tag{3–15}$$

Since the bicarbonate and carbonate concentrations are equal, the hydrogen ion concentration is equal to K_2, and the pH is 10.3.

Accurate calculations of concentrations of species during titrations must include the effect of dilution by the titrant, but thus far those caused by the addition of hydrochloric acid have not been considered. To correct calculations of concentrations of the major components for dilution, multiply each calculated concentration by the factor $V/(V + v)$, where V is the volume of the original solution and v is the volume of hydrochloric acid added at any point. Although in the present example the effect is slight, in many systems the correction is significant.

pH at Point C. The first equivalence point (C in Figure 3-1) is reached when 1 mole of hydrochloric acid per mole of carbonate has been added. This solution contains only sodium bicarbonate; $[H^+]$ is calculated by

$$[H^+] = \sqrt{K_1 K_2} = \sqrt{(3.5 \times 10^{-7})(5 \times 10^{-11})} = 4.2 \times 10^{-9}\ M \tag{3–16}$$

and the pH is 8.4.

pH at Point D. Protonation of half the bicarbonate gives an equimolar solution of bicarbonate and carbonic acid (Point D). This is again a buffer system, this time involving the first dissociation constant of carbonic acid. The calculation is handled in the same way as for Point B, with K_1 used in place of K_2, to yield a pH of 6.5.

pH at Point E. At the second equivalence point (E) the pH is determined by the extent of dissociation of carbonic acid, the principal species present, and [H⁺] is calculated from Equation (3-5):

$$K_1 = 3.5 \times 10^{-7} = \frac{[H^+]\,[HCO_3^-]}{(0.1)\,[50/(50+100)]} = \frac{[H^+]^2}{0.033} \qquad (3-17)$$

Therefore,

$$[H^+] = 1.07 \times 10^{-4}\ M = 10^{-3.97} \qquad (3-18)$$

The pH is 3.97, or, rounding to 2 significant figures, 4.0.

Detection of the Equivalence Point

Either the first or second equivalence point (C or E in Figure 3-1) can be used for carbonate analysis. In neither case is the pH change large in the region of the equivalence point. An uncertainty of 0.1 pH unit at either end point results in an uncertainty of about 1% in the amount of hydrochloric acid required. The error can be reduced if the titration is carried to a preselected indicator color. When a solution is titrated to the second equivalence point, a better approach is to take advantage of the dissociation of carbonic acid into a solution of carbon dioxide in water. Shaking or boiling a solution of carbonic acid causes the equilibrium

$$H_2CO_3 \rightleftarrows H_2O + CO_2\,(g) \qquad (3\text{-}19)$$

to be driven to the right through loss of carbon dioxide. If a carbonate or bicarbonate solution is titrated to just before the equivalence point at pH 4 and then shaken or boiled,[2] the pH will rise to about 8 as the concentration of carbonic acid drops (dotted line in Figure 3-2). The pH is no longer controlled by dissociation of a relatively large concentration of carbonic acid but by a small concentration of bicarbonate. When the titration is continued, the pH goes down sharply because the amount of carbonic acid formed is small and the buffering effect negligible (dashed line in Figure 3-2).

---

[2] In mammals the CO_2 produced through biological oxidation is carried by the blood to the lungs, where it is exchanged for oxygen. Part of the CO_2 is present in the blood as H_2CO_3. Since the time available in the lungs for exchange is short, the dissociation of H_2CO_3 to CO_2 and H_2O is accelerated by the enzyme carbonic acid anhydrase, a zinc-containing protein of high molecular weight. Thus nature need not resort to either boiling or shaking.

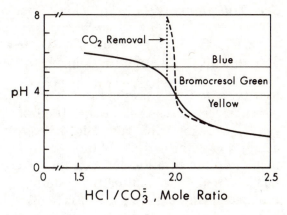

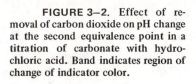

FIGURE 3–2. Effect of removal of carbon dioxide on pH change at the second equivalence point in a titration of carbonate with hydrochloric acid. Band indicates region of change of indicator color.

Standard Solutions

Some standard solutions can be prepared directly by weighing or measuring carefully a definite quantity of a pure substance, dissolving it in a suitable solvent, and diluting it to a known volume. None of the strong acids, however, is convenient to handle and measure accurately in concentrated form. Therefore a solution of approximately the desired molarity is prepared, and the exact value is determined by standardization against a primary-standard base.

Primary standards are stable, nonhygroscopic substances that react quantitatively and are easy to purify and handle. A high equivalent weight is advantageous because weighing errors are minimized. Among the excellent primary standards available are potassium acid phthalate, benzoic acid, oxalic acid dihydrate, and sulfamic acid for standardizing bases and sodium oxalate, tris(hydroxymethyl)aminomethane, 4-amino pyridine, and sodium carbonate for standardizing acids. Pure anhydrous sodium carbonate, besides having all the properties of a suitable primary-standard base, has the added advantage in this experiment of being the same compound as the substance determined. This tends to compensate for determinate errors in end-point selection.

Procedure (median time 3.5 hr)

Preparation of 0.2 M HCl

Put a little less than 1 liter of distilled water into a clean 1-liter bottle. Calculate the volume of 6 M HCl required to prepare 1 liter of 0.2 M HCl, and measure this quantity into a small graduated cylinder. Transfer it to the bottle and mix thoroughly. Label.

Standardization of HCl with Primary-Standard Na$_2$CO$_3$

Dry 1.5 to 2.0 g of pure Na$_2$CO$_3$ in a glass weighing bottle or vial at 150 to 160°C for at least 2 hr.[3,4] Allow to cool, in a desiccator if necessary, and then weigh by difference (to the nearest 0.1 mg) three or four 0.35- to 0.45-g portions of the dry material into clean 200-ml conical (Erlenmeyer) flasks. Add about 50 ml of distilled water to each and swirl gently to dissolve the salt. Add 4 drops of bromocresol green indicator and titrate with the HCl solution to an intermediate green color. At this point stop the titration and boil the solution gently for a minute or two, taking care that no solution is lost during the process. Cool the solution to room temperature, wash the flask walls with distilled water from a wash bottle, and then continue the titration to the first appearance of yellow. Just before the end point the titrant is best added in fractions of a drop.[5] Record the buret reading and add to it the buret calibration correction.

Calculate the molarity of the HCl solution. The procedure outlined in the discussion of calculations below may be used as a guide. Relative deviations of individual values from the average should not exceed about 2 parts per 1000.

Determination of Carbonate in a Sample

Dry the sample in a weighing bottle or vial at least 2 hr at 150 to 160°C. Weigh into clean 200-ml conical flasks, to the nearest 0.1 mg, samples of 0.35 to 0.45 g of the dry material. Dissolve the samples and titrate as in the standardization procedure.

Calculate and report the percentage of Na$_2$CO$_3$ in the sample. Use the Q test (Section 2-1) as the criterion for rejection of suspect experimental data. Either the median or the average may be reported. When the median is chosen the median value for the molarity of the HCl should be used in the calculations rather than the average value. See Section 2-1 for a discussion of medians and averages.

---

[3] Na$_2$CO$_3$ tends to absorb H$_2$O from the air to form Na$_2$CO$_3$·H$_2$O, and CO$_2$ to form NaHCO$_3$. At least several hours of drying at 140°C is necessary to remove all H$_2$O and CO$_2$.

[4] Use a pencil or felt marking pen to label the container with the name or sample number of the contents and with your locker number. The container may be placed inside a small glass beaker, and a watch glass, raised with several bent portions of glass rod, placed on top for protection. Avoid leaving chemicals or equipment in the drying oven longer than necessary; this not only causes crowding, but increases the chance of equipment being broken or samples contaminated by spilled chemicals.

[5] To deliver amounts less than 1 drop from a buret, first let a droplet form on the tip, and then touch the tip momentarily to the inside wall of the flask. Rinse the wall with a small amount of distilled water from a wash bottle to ensure that the titrant is washed into the solution. Do not rinse the tip of the buret.

Calculations

The percentage of Na_2CO_3 in a sample can be calculated in two steps: (1) the determination of the molarity of the HCl titrant from the standardization titrations and (2) the calculation of the percentage of Na_2CO_3 from titrations of the sample.

1. **Molarity of HCl.** In titrations of Na_2CO_3 with HCl to the pH 4 end point, 2 moles of HCl are added for each mole of Na_2CO_3:

$$2HCl + Na_2CO_3 \rightleftarrows H_2CO_3 + 2NaCl \qquad (3–20)$$

The HCl molarity is obtained from the following relations:

$$M_{HCl} = \frac{\text{moles HCl}}{\text{liter}} = \frac{(\text{moles } Na_2CO_3)\,(2)}{(\text{ml HCl}/1000)}$$

$$= \frac{(\text{g } Na_2CO_3)\,(2)}{(\text{mol wt } Na_2CO_3)\,(\text{ml HCl}/1000)} \qquad (3–21)$$

The factor 2 is required because each mole of Na_2CO_3 reacts quantitatively with 2 moles of HCl.

2. **Percentage of Na_2CO_3 in Sample.** The percentage of Na_2CO_3 in the sample is calculated as follows:

$$\% \, Na_2CO_3 = \frac{\text{g } Na_2CO_3 \text{ in sample}}{\text{wt of sample}} \, (100)$$

$$= \frac{(\text{moles } Na_2CO_3)\,(\text{g } Na_2CO_3/\text{mole})}{\text{wt of sample}} \, (100) \qquad (3–22)$$

$$= \frac{(\text{ml HCl})\,(\text{molarity HCl})\,(\text{g } Na_2CO_3/\text{mole})}{(1000)\,(2)\,(\text{wt of sample})} \, (100)$$

Remember: Poor results are often caused by errors in calculation rather than by faulty laboratory technique. Check all calculations before reporting results.

Example 3–2.

A 0.3729-g sample of a mixture of Na_2CO_3 and inert material required 32.77 ml of HCl for titration. A 0.4404-g sample of pure Na_2CO_3 required 40.12 ml of the same HCl solution. What is the percentage of Na_2CO_3 in the mixture?

First calculate the HCl molarity:

$$M_{HCl} = \frac{(0.4404 \text{ g}) (2)}{(40.12 \text{ ml}/1000) (105.99 \text{ g/mole})} = 0.2071 \text{ moles/liter}$$

Then calculate the percentage of Na_2CO_3 in the sample:

$$\% \ Na_2CO_3 = \frac{(32.77 \text{ ml}) (0.2071 \text{ moles}/1) (105.99 \text{ g/mole})}{(1000) (2) (0.3729 \text{ g})} (100)$$

$$= 96.45\%$$

Notice that the units cancel out.

PROBLEMS

3–9.   A 0.3752-g sample of pure sodium carbonate required 38.21 ml of a hydrochloric acid solution to reach the bromocresol green end point. What is the molarity of the hydrochloric acid?

3–10.   A 0.5063-g sample of a mixture of sodium and potassium carbonates required 41.84 ml of a 0.2023 M hydrochloric acid solution to reach the bromocresol green end point. What is the percentage of carbonate as $CO_3^=$ in the sample?

3–11.   Which pairs of the four compounds sodium hydroxide, sodium bicarbonate, sodium carbonate, and carbonic acid can exist in significant concentrations in solution?

3–12.   Sketch the approximate titration curves expected (calculations not required) for the titration with 0.1 M hydrochloric acid of (a) a pure sample of sodium bicarbonate; (b) an equimolar mixture of sodium bicarbonate and sodium carbonate; (c) an equimolar mixture of sodium hydroxide and sodium carbonate. Label the axes clearly and completely.

3–13.   A slurry contains two of the compounds sodium hydroxide, sodium bicarbonate, and sodium carbonate. A 0.2116-g sample requires 36.55 ml of 0.1020 M hydrochloric acid to reach a phenolphthalein end point (pH 8.4) and an additional 6.34 ml to reach the bromocresol green end point. Which compounds are present, and what is the percentage of each?

3–14.   Calculate the pH at the start, midpoint, and equivalence point for the titration of 0.1 *M* ammonia with 0.1 *M* hydrochloric acid. Sketch the titration curve expected on the basis of the calculations. Select a suitable indicator from the list in Appendix C.

3–15.   Why can borax, a sodium salt of boric acid, be titrated precisely by an acid-base reaction, whereas boric acid cannot?

3–16.   A sample contains pyridine hydrochloride, a salt of pyridine. Could a satisfactory acid-base titration of this material be obtained with a visual indicator? If so, explain why, naming a suitable titrant and indicator.

3–17.   What fraction of the acid-base indicator phenol red is in the acidic form at pH 6 and pH 8?

3–18.   What volume of 37% hydrochloric acid of density 1.18 would be required to prepare 2 liters of 0.30 *M* solution?

SELECTED REFERENCES

I. M. Kolthoff, E. B. Sandell, E. J. Meehan, and S. Bruckenstein, *Textbook of Quantitative Inorganic Analysis,* 4th ed., Macmillan, New York, 1969, p 778.
D. G. Peters, J. Hayes, and G. Hieftje, *Chemical Separations and Measurements,* W. B. Saunders Company, Philadelphia, 1974, Chapter 4.
A. L. Underwood, *Anal. Chem.* **33,** 955 (1961). The enzyme carbonic acid anhydrase accelerates the hydration of carbon dioxide dissolved in water; the resulting carbonic acid can be titrated rapidly and accurately.

3–3   COMPLEXATION TITRATIONS: DETERMINATION OF CALCIUM WITH EDTA

> Probably never in the history of chemistry has a method been adopted so quickly and universally as the EDTA titration for the hardness of water. In fairness, it should be added that never in history has there been a method so poor as the soap titration it displaced.
>
> *H. Diehl*

Background

A complex is an ion or a molecule formed by a reaction between two ions or molecules capable of independent existence. The most important complexation reactions from an analytical point of view are those in which one of the reactants is a metal ion. The reaction of a metal ion of charge n with a singly charged ligand, L^-, can be written

$$M^{n+} + xL^- \rightleftharpoons ML_x^{\,n-x} \tag{3-23}$$

The equilibrium constant for this reaction[1] is

$$K_{eq} = \frac{[ML_x^{\,n-x}]}{[M^{n+}]\,[L^-]^x} \tag{3-24}$$

If the equilibrium constant is sufficiently large, the reaction may be analytically useful, provided interferences are absent or eliminated and a method of end-point detection is available. Also, for quantitative results, if more than one coordinating species or ligand can combine with a metal, the overall formation constant must be large enough to allow the final complex to form completely. Since individual stepwise formation constants are frequently small and similar in magnitude, reagents forming 1:1 complexes with metals are preferred as titrants.

Rates of formation or dissociation of metal complexes are usually fast. A few ions, such as cobalt (III) and chromium (III) in the first row of transition elements, form complexes too slowly to be determined by direct complexation titration. Such ions may be analyzed by adding a known excess of complexing agent, allowing the reaction to go to completion, and back titrating the excess with a standard solution of a metal ion.

Chemistry of EDTA

The most important complexing reagent in analytical chemistry is ethylenediaminetetraacetic acid (EDTA), often written H_4Y. The tetrabasic anion of this acid, Y^{4-},

forms complexes with virtually all metal ions (Figure 3-3). Even sodium ions are complexed to some extent. The metal atom is usually held in a

---

[1] Equilibrium constants in complexation reactions are frequently written in this way and are called formation or stability constants. This practice contrasts with that in acid-base equilibria, where dissociation constants (the reciprocals of formation constants) are more common.

Li	Be												B	C	N	O	F
Na	Mg												Al	Si	P	S	Cl
K	Ca	Sc	Ti	V	Cr	Mn	Fe	Co	Ni	Cu	Zn	Ga	Ge	As	Se	Br	
Rb	Sr	Y	Zr	Nb	Mo			Pd	Ag	Cd	In	Sn	Sb		I		
Cs	Ba		Hf	Ta	W	Re			Au	Hg	Tl	Pb	Bi				

| La Series | La | Ce | Pr | Nd | | Sm | Eu | Gd | Tb | Dy | Ho | Er | Tm | Yb | Lu |
| Ac Series | | Th | | U | | Pu | | | | | | | | | |

FIGURE 3–3. Elements determinable in some form by titration with EDTA. Shaded elements are determinable by indirect methods.

1:1 complex having four or five 5-membered rings (Figure 3-4). Such ring structures, formed by metal coordination to two or more groups on the same ligand, are called chelate rings. Ring formation contributes appreciably to the stability of the complexes.

The pK_a values (negative logarithms of the acid dissociation constants) for EDTA are 2.2, 2.7, 6.2, and 10.0. It can be seen from these values that the first two protons are strongly, the third weakly, and the fourth very weakly acidic. Since most metals form complexes only with the tetraanion form, Y^{4-}, metal ions must compete with hydrogen ions for coordination sites on the ligand. Consequently, only ions that have

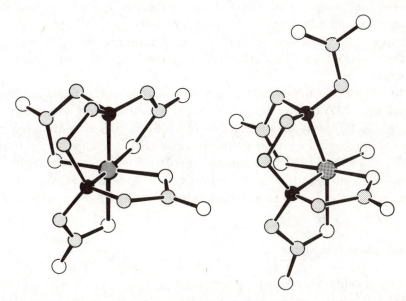

FIGURE 3–4. Crystal structures of cobalt (left) and nickel (right) complexes with EDTA. The nickel structure has an oxygen atom from a water molecule replacing one oxygen of EDTA in the sixth coordination position of the nickel. Heavy shading, metal; solid, nitrogen; light shading, carbon; no shading, oxygen. Hydrogen atoms are not shown.

TABLE 3–1. SOME METAL ION-EDTA FORMATION
CONSTANTS

(For the reaction $M^{n+} + Y^{4-} \rightleftarrows MY^{n-4}$)

Metal Ion	log K_{MY}	Metal Ion	log K_{MY}	Metal Ion	log K_{MY}
Na^+	1.7	La^{3+}	15.2	Pd^{++}	18.5
Li^+	2.8	Ce^{3+}	15.8	Ni^{++}	18.6
Ag^+	7.3	Al^{3+}	16.1	Cu^{++}	18.8
Ba^{++}	7.8	Co^{++}	16.2	Hg^{++}	22.1
Sr^{++}	8.7	Zn^{++}	16.5	Th^{4+}	23.2
Mg^{++}	8.7	Cd^{++}	16.5	In^{3+}	25.0
Ca^{++}	10.7	Ti^{3+}	17.7	Fe^{3+}	25.1
Mn^{++}	13.8	Pb^{++}	17.9	Bi^{3+}	27.9
Fe^{++}	14.3	Pu^{3+}	18.1	Co^{3+}	40.7

formation constants greater than about 10^{15} form EDTA complexes in acidic solution. A list of metal formation constants with EDTA is given in Table 3-1. It is apparent that EDTA is not a selective titrant, because it complexes strongly with nearly all the metals listed. Nevertheless, by pH adjustment advantage can be taken of differences in formation constants. Also, masking agents can be used to block the reaction of certain metals. For example, lead can be titrated with EDTA in the presence of zinc if cyanide is added to complex the zinc, or in the presence of aluminum if triethanolamine is added to chelate the aluminum. By such techniques specific titrations can be performed on many mixtures without preliminary separation.

In this experiment a mixture of calcium and magnesium salts is titrated with EDTA. The determination of such mixtures is important; for example, sulfate and bicarbonate salts of these two ions are responsible for water hardness. From the formation constants for these metal ions and the pK_a values for EDTA, it can be calculated that the pH must be 10 or higher for complexation to be quantitative. A buffer of ammonia and ammonium chloride serves well in this pH region (pK_b for NH_3 is 4.75).

Complexation Indicators

The development of a series of visual indicators covering a wide range of metal formation constants has been an important factor in the widespread application of EDTA titrations. One of the best indicators for calcium and magnesium is Calmagite, which at pH 10 forms a complex with magnesium that has a distinct red color at magnesium concentrations as low as $10^{-6}\,M$. Figure 3-5 shows the structure and colors of the magnesium complex and of the protonated forms of Calmagite. The indicator is a tribasic acid, the sulfonic acid proton being completely dissociated in water and the two hydroxyl protons having pK_a values of 8.1 and 12.4. The log of the formation constant for the reaction

$$Mg^{++} + HIn^= \rightleftarrows MgIn^- + H^+ \qquad (3-25)$$

at pH 10 is 5.7. Note that the presence of hydrogen ion in the reaction makes it pH dependent. An equilibrium constant for a reaction that depends on conditions present in solution, such as pH, is called a *conditional constant*.

In a titration Calmagite works in the following way. When a small amount is added to a solution containing magnesium at pH 10, red MgIn⁻ forms. As EDTA titrant is added, free magnesium ions in solution react with it. After all the free magnesium is complexed, the next portion of

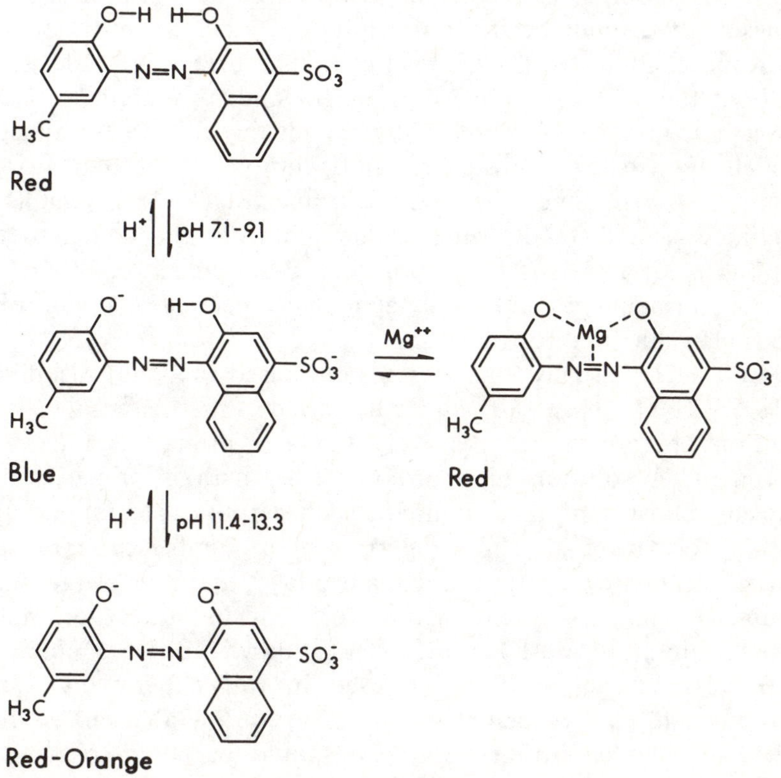

FIGURE 3-5. Structure of the indicator Calmagite at several pH values and of the magnesium-Calmagite complex at pH 10.

EDTA removes magnesium from the indicator complex, converting the indicator to the blue $HIn^=$ form:

$$MgIn^- + HY^{3-} \rightleftarrows MgY^= + HIn^= \qquad (3-26)$$
$$\text{(red)} \qquad\qquad\qquad \text{(blue)}$$

At pH values above 12 or below 8 the free indicator is red, like the magnesium complex, and so a color change is not observed.

The calcium-Calmagite complex is too weak to function as an indicator (log K_{CaIn^-} is 3.7 at pH 10). Calcium can, however, be titrated if a trace of magnesium is present. Magnesium can be added to either the titrant or the solution being titrated. When added to the solution being titrated, the amount of magnesium added must be known; when added to the titrant before standardization, its presence is accounted for in the standardization step. The latter method is employed in this experiment because the primary standard, pure calcium carbonate, contains no magnesium.

The optimum pH range for titrations of calcium and magnesium with EDTA is about 9 to 11. In solutions more acidic than pH 9, formation of metal-EDTA complexes is incomplete, whereas in solutions above pH 12, magnesium hydroxide begins to precipitate.

Another indicator for EDTA titrations is Eriochrome Black T, which has about the same color-change properties and applicability as Calmagite. Although it may be used essentially interchangeably with Calmagite, the color change is slightly less sharp, and solutions of the indicator tend to decompose with time. Solid Eriochrome Black T is stable and is sometimes ground 1:100 with sodium chloride and added to titration solutions as a powder. A large number of other indicators that operate on the same principle as does Calmagite have been developed for use at various pH values and for various metals.

EDTA is usually purchased and used as the disodium salt, $Na_2H_2Y \cdot 2H_2O$. This salt must be dried under carefully controlled conditions before it can be weighed as a primary standard in this experiment. A solution of approximate strength is prepared and standardized against a primary standard such as pure calcium carbonate. In addition to having all the properties of a suitable primary standard, calcium carbonate has the added advantage that the element titrated, calcium, is a major element in limestone and in hard water. Magnesium iodate tetrahydrate, $Mg(IO_3)_2 \cdot 4H_2O$, is another suitable standard.

Limestone samples may be used in this experiment. Limestone dissolves readily in hydrochloric acid. Since considerable carbon dioxide is evolved in the dissolution step, care should be taken to avoid loss of sample in the form of spray. Most limestones contain varying amounts of iron and aluminum. Because these metals form complexes with Calmagite more stable than the magnesium complex, they block proper indicator action. Aluminum can be kept from interfering by the addition of triethanolamine, which forms a more stable chelate with aluminum than does Calmagite. This operation is called masking. Iron is masked by cyanide, but because cyanide forms a complex with iron(III) slowly, ascorbic acid is added to reduce the iron to iron(II) prior to addition of cyanide.

A Refinement in Technique: Alternation

Alternation consists of carrying out a series of operations on standards and samples in alternating sequence. In this way errors resulting from changes in solution concentrations, instrument drift, end-point judgment, and so forth, tend to be canceled. Competent analytical chemists use this technique whenever possible, even though it is not written into the procedures. For this experiment alternation involves preparation for titration of both the calcium carbonate standards and the samples. The titrations are then done in the following order: standard, sample, standard, sample, and so on. Procedural modifications needed to incorporate alternation are generally obvious and usually are not included in the directions. This technique requires only slightly more organization and adds polish to experimental work.

Planning Experimental Work

At this point look ahead and plan the experimental work for the next few laboratory periods. Read the next experiment or two, and organize the work to make best use of your time.

Procedure (median time 3.9 hr)

Preparation and Standardization of EDTA Solution

Prepare an approximately 0.015 M EDTA solution by adding 4 g of the disodium salt of EDTA, 20 ml of 1% $MgCl_2$ solution, and 3 ml of 6 M NH_3 to about 750 ml of water.[2] In this experiment use only distilled water purified by passage through a column containing either a cation-exchange resin or a mixture of cation- and anion-exchange resins. Prepare a buffer solution of 3 g NH_4Cl in 50 ml of 6M NH_3.

---

[2] The disodium salt of EDTA, $Na_2H_2Y \cdot 2H_2O$, is the most commonly used form, being easier to prepare commercially and more soluble in water than the acid, H_4Y. It is somewhat slow to go into solution, however, requiring 10 to 15 min for complete dissolution. Occasional shaking speeds the process, as does gentle warming.

Weigh to the nearest 0.1 mg approximately 0.5 g of dry $CaCO_3$ on a sheet of glazed weighing paper. Transfer quantitatively to a funnel inserted in the top of a 100-ml volumetric flask, using a camel-hair brush to aid in transfer of the last particles. Rinse the material into the flask with a ml or two of 12 M HCl, then add 10 more ml of 12 M HCl by rapid dropwise addition.[3] Warm if necessary to complete the dissolution.[4] When dissolution is complete, remove the funnel, rinsing inside and out with deionized water from a wash bottle. Fill the flask to the mark and mix well. Using a calibrated 10-ml pipet, measure aliquots of the standard calcium solution into each of three or four 200-ml conical flasks. Immediately before each titration add 50 mg of ascorbic acid, 10 ml of NH_3-NH_4Cl buffer, and 10 ml of a solution 0.4 M in NaCN (*Caution: Poison*) and 0.75 M in triethanolamine, mixing after each addition.[5] Then add 4 to 5 drops of Calmagite indicator solution. Titrate with EDTA solution until the indicator changes from red to pure sky blue with no tint of red.

Titration of Sample

In a 100-ml beaker weigh 0.5 g of sample to the nearest 0.1 mg. Dissolve as for the standard.[6] A small residue of insoluble silicate minerals may be ignored. Dilute to 100 ml in a volumetric flask, pipet 10-ml aliquots into 200-ml flasks, and titrate in the same way as the standard.[7]

Report the percentage of calcium in the sample.[8]

Determination of Water Hardness

The hardness of natural water may be determined by the following method. Measure 100-ml samples of the water into 200-ml conical flasks.[9]

---

[3] Alternatively the sample may be weighed into a 100-ml beaker and dissolved before transfer. Cover the beaker with a watch glass during the acid addition to avoid loss of spray, and swirl the solution gently after each addition. When dissolution is complete, rinse the undersurface of the watch glass into the beaker with a little deionized water from a wash bottle before quantitative transfer of the solution to the volumetric flask.

[4] Heating of volumetric flasks to moderate temperatures on a hot plate does not affect calibration, as the original volume is regained upon cooling to room temperature.

[5] The cyanide solution may be dispensed by the laboratory instructor. Follow the directions for disposal carefully, as contact with acid produces poisonous HCN gas.

[6] For limestone samples, heating for ½ to 1 hr at 80 to 90° usually is necessary for dissolution.

[7] Most difficulties with indistinct end points arise from incorrect pH adjustment.

[8] The result could be reported in a number of ways, such as % $CaCO_3$, % $MgCO_3$, or % Mg. Here it is reported on the basis that all the material titrated is calcium. This is common reporting procedure.

[9] Students having access to water other than the local supply may wish to bring samples for titration. Otherwise, use tap water. If the water is unusually hard, 50-ml samples may suffice. A graduated cylinder is adequate for measurement of these samples.

Prepare a pH 10 buffer by dissolving 0.35 g of NH_4Cl and 7 ml of 6 M NH_3 in water and diluting to 25 ml. Add 5 ml of buffer solution and 4 to 5 drops of indicator to each flask and titrate as outlined above.

Calculations

The concentration of the EDTA solution is found by

$$\text{molarity of EDTA} = \frac{(\text{g CaCO}_3)(V_{10}/100)}{(\text{mol wt CaCO}_3)(\text{ml EDTA}/1000)}$$

where V_{10} is the calibrated volume of the 10-ml pipet. The percentage of calcium in the sample is then calculated by

$$\% \text{ Ca} = \text{wt Ca}(100)/\text{wt sample}$$

$$\text{wt Ca} = (\text{ml EDTA}/1000)(\text{molarity EDTA})(\text{mol wt Ca})$$

$$\text{wt sample} = (\text{initial wt of unknown})(V_{10}/100)$$

Water hardness is usually expressed not in percentage of calcium as for the limestone sample but in parts per million (ppm), or milligrams per liter, of $CaCO_3$:

$$\text{hardness} = \frac{(\text{ml EDTA}/1000)(\text{molarity EDTA})(\text{mol wt CaCO}_3)(1000)}{\text{volume of sample in liters}}$$

PROBLEMS

3–19.   A limestone sample weighing 0.8574 g was dissolved in hydrochloric acid and the solution diluted to 100.0 ml. A 10.00-ml sample of this solution required 36.20 ml of 0.02027 M EDTA for titration. What was the percentage of calcium carbonate in the sample?

3–20.   A 100.0-ml sample of river water required 4.61 ml of 0.01619 M EDTA for titration by the procedure of this experiment. What was the hardness of the water in milligrams of calcium carbonate per liter, in parts per million of calcium carbonate, and in milligrams of magnesium carbonate per liter?

3–21.   What volume of 6 M NH_3 should be added to 100 ml of a solution containing 8 ml of 12 M HCl to bring the pH to 10?

3–22.   Why do the directions for this experiment specify water that has been passed through a cation-exchange resin?

3–23.†   If the conditional formation constant of the magnesium complex with Calmagite at pH 10 is $10^{5.7}$, what is the true formation constant, that is, the constant under conditions where Calmagite is in the In^{3-} form? [Use K_3 for Calmagite together with K_{MY} for $MgY^=$ from Table 3-1 and Equation (3-25)].

SELECTED REFERENCES

A. Ringbom, *Complexation in Analytical Chemistry,* Interscience, New York, 1963. General reference for metal titrations with EDTA.

F. Lindstrom and H. Diehl, *Anal. Chem.* **32**, 1123 (1960). Introduction of Calmagite indicator.

A. E. Martell and L. G. Sillen, *Stability Constants of Metal-Ion Complexes,* 2nd ed. The Chemical Society, London, Special Publication 17, 1964. Supplement 1, 1971. A compilation of stability constants for both organic and inorganic ligands with metal ions and with hydrogen ion. Includes methods and conditions of measurement along with literature references.

D. G. Peters, J. Hayes, and G. Hieftje, *Chemical Separations and Measurements,* W. B. Saunders Company, Philadelphia, 1974, Chapter 6.

Chapter 4

TITRIMETRIC ANALYSIS: OXIDATION–REDUCTION

4–1 GENERAL BACKGROUND

This chapter concerns some of the practical aspects of oxidation-reduction reactions in volumetric analysis. An understanding of the basic theoretical principles is assumed. Four experimental procedures are provided: iron in an ore by dichromate and by permanganate, copper by iodimetry, and ethylene glycol by periodate.

Whether a chemical electron-transfer process can be used for analysis depends in part on the equilibrium constant for the overall process being sufficiently large. This constant can be calculated from the standard electrode potentials of the half-reactions involved. A considerable difference between calculated and actual equilibrium constants may exist, however, because solution conditions during a titration are seldom those under which standard potentials are measured. Another consideration is whether the reaction proceeds rapidly enough to be analytically useful. Still other requirements are that the species being determined not be involved in side reactions and that other substances in the sample not interfere.

The experimental procedure used in volumetric oxidation-reduction analyses consists of (1) preparation and standardization of the titrant, (2) sample preparation, including dissolution and preliminary adjustment of the oxidation state of the substance to be determined, and (3) titration. The principal oxidizing titrants are potassium permanganate, potassium dichromate, cerium(IV), iodine, and potassium bromate. The principal reducing titrants are sodium thiosulfate, iron(II), titanium(III), and chromium(II). Useful primary standards include arsenic trioxide, potassium iodide, sodium oxalate, potassium dichromate, copper, potassium iodate, potassium bromate, and potassium ferrocyanide.

Preliminary adjustment of the oxidation state of a sample generally is accomplished by adding an excess of a reagent whose electrode potential is high (or low) enough to effect the required oxidation (or reduction). Occasionally electrochemical methods are employed. A method of removing the excess preliminary oxidant or reductant usually is needed to avoid interference in the subsequent titration. Several preoxidants and prereductants are listed in Table 4-1.

End points in redox titrations can be detected by either visual indicators or instrumental methods. Visual indicators may be general or specific. General indicators change color when the solution potential reaches a certain value, whereas specific ones respond only to a particular chemical system. Ferroin is an example of the general type, and starch (for iodine titrations) an example of the specific. Instrumental methods using potentiometry are described in Experiments 8-2 and 8-3.

TABLE 4–1. SOME REAGENTS USED FOR PRELIMINARY OXIDATION OR REDUCTION IN ANALYTICAL OXIDATION-REDUCTION TITRATIONS

Reagent	Approximate Electrode Potential[a]	Method of Removing Excess Before Titration
Preoxidants		
Hot 72% perchloric acid, $HClO_4$[b]	2.0	Dilution with water and cooling; free Cl_2 removed by boiling
Potassium persulfate, $K_2S_2O_8$	2.0	Boiling; usually used with Ag^+ catalyst
30% Hydrogen peroxide, H_2O_2	1.8	Boiling; used in alkaline solution
Potassium periodate, KIO_4	1.7	Precipitation as Hg_5IO_6; most effective in hot solutions
Sodium bismuthate, $NaBiO_3$	1.6	Filtration
Prereductants		
Bismuth amalgam, BiHg	0.3	Liquid–liquid separation
Silver metal in HCl solution, Ag	0.2	Filtration; often used in column form (Walden reductor)
Tin(II) chloride, $SnCl_2$	0.1	Oxidation with $HgCl_2$ to give tin(IV) and Hg_2Cl_2
Sulfur dioxide, SO_2	0.1	Boiling
Zinc metal, Zn	−0.8	Filtration; often used in column form (Jones reductor)

[a] The presence of appreciable amounts of products is assumed.

[b] Perchloric acid must be used only under controlled conditions because of the danger that the hot, concentrated acid may react explosively with oxidizable material. The fumes should not be allowed to condense on surfaces where they can combine with organic material to form shock-sensitive products.

4–2 DICHROMATE DETERMINATION OF IRON IN AN ORE

Background

In this experiment a sample of iron ore is analyzed for iron by titration of iron(II) with a solution of potassium dichromate. Iron(III) must be reduced to iron(II) before titration; this reduction is carried out here with tin(II) chloride.

Potassium dichromate has a lower electrode potential than many other oxidizing titrants, and its rate of reaction with many materials is slow. Nevertheless, it possesses several useful properties. Since the solid is easy to obtain in pure form, the dry salt can be used directly as a primary standard. Solutions of dichromate are stable indefinitely and may even be boiled without decomposition. The lower electrode potential is in some instances an advantage; for example, in the titration of iron(II), chloride is not oxidized unless its concentration exceeds about 1 M. For this reason dichromate is the titrant of choice in almost all volumetric determinations of iron in ores, steels, and slags.

Iron, the second most abundant metal in the earth's crust, is the most important industrial element. Although many minerals contain iron, the important ore deposits consist of iron(III) oxide (hematite). Typical samples in this experiment have compositions similar to commercial hematite ores. For analysis, ore samples are first treated with concentrated hydrochloric acid. The iron oxides usually dissolve completely, leaving some insoluble residue, primarily silica. Occasional particles of the black mineral magnetite, Fe_3O_4, also will not dissolve. If an ore contains interfering substances such as titanium, conditions must be adjusted so that these interferences are minimized. This adjustment sometimes can be accomplished without physical separation; in this experiment tin(II) is used as a selective reductant. Titanium(IV), not being reduced by tin(II), does not interfere in the subsequent titration.

The reduction of iron(III) to iron(II) with tin(II) chloride is carried out in hot hydrochloric acid solution. The temperature should be 70 to 90° C to ensure rapid, complete reduction. In concentrated hydrochloric acid, iron(III) forms a series of yellow chloride complexes; the disappearance of the yellow color can be used to indicate when reduction is complete. The sample then is cooled, and the slight excess of tin(II) chloride destroyed with mercury(II) chloride. The mercury(II) chloride must be added rapidly; otherwise it may be reduced to finely divided metallic mercury by local excesses of tin(II). Metallic mercury may be oxidized by dichromate in the subsequent titration and thereby give high results.

Once the reduction of iron is complete, sulfuric and phosphoric acids are added. Sulfuric acid lowers the pH so that the dichromate electrode potential is maintained at a high value. Phosphoric acid serves two purposes. First, it forms a colorless phosphate complex, $FeHPO_4^+$, with iron(III) that prevents yellow iron(III) chloride complexes from interfering with the dichromate end point. Second, the formation of phosphate complexes decreases the electrode potential of the iron(III)-(II) couple, thereby shifting the iron(II)-dichromate equilibrium to the right and resulting in more complete reaction. Also, the decreased potential of the iron couple causes the end point of the indicator, diphenylamine sulfonic acid, to be sharper and to coincide more closely with the equivalence point.

Diphenylamine sulfonic acid is colorless in its reduced form and violet in its oxidized form. The transition potential for the color change is 0.83 V. In the dichromate titration of iron the solution before the end point is green owing to the presence of chromium(III), and so the end-point color change is from green to purple. Although the reaction between dichromate and diphenylamine sulfonic acid is slow in pure solution, catalysis by iron(II) accelerates the rate greatly, and the color transition is rapid and sharp. The indicator reaction is a two-electron oxidation to an intensely purple species that cannot be isolated, but slowly decomposes. This results in a gradual fading of the end-point color, but causes no difficulty. The indicator blank is of the order of 0.04 ml of dichromate titrant.

Even though the reaction between dichromate and iron(II) involves several transient intermediate oxidation states of chromium, the overall process is rapid. The reaction mechanism is not completely understood. (See reference by Laitinen and Harris at end of section for detailed discussion.) From a table of standard electrode potentials (see Appendix C) it can be seen that the oxidation of iron(II) by dichromate should be quantitative. Chloride is not oxidized by dichromate as long as the chloride concentration does not exceed about 1 M.

In summary, the reactions for the dichromate determination of iron in an ore are:

dissolution,

$$Fe_2O_3 + 6HCl \rightarrow 2FeCl_3 + 3H_2O \tag{4-1}$$

reduction,

$$2Fe^{3+} + Sn^{++} \rightleftarrows 2Fe^{++} + Sn^{4+} \tag{4-2}$$

removal of excess tin(II) after iron(III) reduction,

$$Sn^{++} + 2HgCl_2 \text{ (excess)} \rightleftarrows Sn^{4+} + Hg_2Cl_{2(s)} + 2Cl^- \tag{4-3}$$

and titration,

$$6Fe^{++} + Cr_2O_7^{=} + 14H^+ \rightarrow 6Fe^{3+} + 2Cr^{3+} + 7H_2O \qquad (4–4)$$

Tin(II) chloride is commonly chosen for the prereduction step because the standard electrode potential of the tin(IV)-(II) couple is not low enough to reduce the titanium(IV) often present in iron ores. Therefore titanium is not titrated by dichromate and consequently does not interfere.

Another common reductant for sample treatment prior to analysis is zinc metal. Not being selective, it reduces many metals besides iron to low oxidation states where they can be determined by dichromate titration. For example, after zinc reduction, vanadium can be titrated from oxidation state 2 to 4, chromium from 2 to 3, uranium from 4 to 6, titanium from 3 to 4, and molybdenum from 3 to 6.

Procedure (median time 5.5 hr)

Preparation of 0.02 M Dichromate Solution

Weigh accurately 2.6 g of reagent-grade $K_2Cr_2O_7$ in a weighing bottle. Transfer most of the salt into a 500-ml volumetric flask and weigh the bottle plus residual salt. Dissolve in distilled water, dilute to the mark, and mix well.

Dissolution of Iron Standard

Weigh, to the nearest 0.1 mg, 0.2 to 0.25 g of electrolytic iron into 500-ml conical flasks.[1] Add 10 ml of 12 M HCl to each (in a fume hood), cover with a small watch glass, and warm on a hot plate until dissolved. Add HCl, not exceeding 5 to 10 ml, as needed to replace that lost by evaporation. Prereduce the iron to iron(II) by the procedure outlined below.

Prereduction of Iron(III)

Prepare a tin(II) chloride solution by dissolving about 5 g of $SnCl_2 \cdot 2H_2O$ in 35 ml of 6 M HCl. As this solution is readily air oxidized, keep it covered, and discard it at the end of the laboratory period.

From this point each sample must be carried separately through the

---

[1] Iron wire also may be used as a standard, though it varies in quality. Since it is also prone to rust, it should be polished with fine emery cloth and a clean towel before use.

remainder of the procedure. Heat the sample to nearly boiling. Remove from heat, and slowly add tin(II) chloride solution dropwise until the orange-yellow color almost disappears. Rinse the sides of the flask with a small amount of water. Again heat to nearly boiling and continue dropwise addition of tin(II) chloride solution until the yellow or yellow-green disappears.[2] Add 2 drops excess; any more may result in formation of metallic mercury instead of Hg_2Cl_2 in the next step. If through accident more than 2 drops of excess tin(II) chloride is added, or if there is doubt as to the completeness of iron(III) reduction, add a few drops of $KMnO_4$ solution until the yellow-green color returns and decolorize again. Immediately cool the sample to room temperature by holding it under a stream of cold water. Add 100 ml of freshly boiled and cooled water and then 10 ml of saturated mercury(II) chloride solution. (*Caution*: $HgCl_2$ is poisonous.) A white silky precipitate should form. If no precipitate is visible, or if a dark precipitate forms, discard the sample and take more care with subsequent ones.[3]

Allow the solution to stand 3 to 5 min.[4]

Titration of Iron(II) with Dichromate

Add 10 ml of 3 M H_2SO_4, 20 ml of 6 M H_3PO_4, and 4 to 6 drops of sodium diphenylamine sulfonate indicator solution. Titrate with $K_2Cr_2O_7$ solution, swirling constantly and adding the titrant slowly near the end point. The end point is the first appearance of purple in the gray-green solution.

Procedure for Dissolution of Sample

Accurately weigh samples of iron ore into 500-ml conical flasks.[5] Add 10 ml of concentrated HCl to each in a fume hood, cover with a small watch glass, and warm on a hot plate until the iron oxide portion of the sample dissolves (about 1 hr).[6] Do not allow the samples to evaporate

---

[2] The color of the solution at this point will depend on the nature and origin of the sample. In any case, the solution will change abruptly from a pronounced to a faint yellow within 1 to 2 drops of $SnCl_2$, indicating complete reduction of the iron(III).

[3] If no Hg_2Cl_2 precipitate forms, the amount of $SnCl_2$ added was insufficient, and the iron is probably not all reduced. A dark precipitate indicates the presence of finely divided mercury metal, the result of too much $SnCl_2$.

[4] The reaction between tin(II) and mercury(II) requires several minutes to go to completion. A long delay, however, may result in appreciable air oxidation of the iron(II).

[5] The weight required for samples containing on the order of 60% Fe_2O_3 is about 0.4 g. Consult the laboratory instructor for the range of composition of the samples provided.

[6] The iron oxide in the samples is dissolved when no dark specks of sample remain. A white or gray-white residue is silica and may be ignored. Dissolution may be hastened, if desired, by gentle boiling with frequent swirling.

to dryness, or iron oxide may bake onto the glass and be difficult to redissolve. Add HCl if necessary to replace that lost by evaporation.

Proceed with the reduction by tin(II) chloride and titration as outlined above.

Report the percentage of iron in the sample.

Calculations

Calculate the molarity of the potassium dichromate solution from the weight of dichromate taken and also from the standardization against pure iron metal. The two values should agree within 2 to 3 parts in 1000.

In the standardization against iron

$$\text{moles } K_2Cr_2O_7 \text{ required} = (1/6)(\text{moles Fe taken})$$

Then

$$M\ K_2Cr_2O_7 = \frac{\text{moles } K_2Cr_2O_7}{\text{ml } K_2Cr_2O_7/1000} = \frac{\text{wt Fe}}{(\text{wt Fe/mole})(6)(\text{ml } K_2Cr_2O_7/1000)}$$

Before reporting results for this determination, check the calculation procedure by working the problems below.

PROBLEMS

4–1.   A dichromate solution was prepared by weighing 2.488 g of potassium dichromate into a 500-ml volumetric flask, dissolving, and diluting to 500.0 ml with water. What was the molarity of the resulting solution?

4–2.   A solution of potassium dichromate was standardized against iron metal by the procedure of this experiment. If 0.2317 g of iron required 38.72 ml, what was the molarity of the dichromate solution?

4–3.   A 0.5069-g sample of iron ore, analyzed by the procedure of this experiment, required 43.36 ml of 0.01711 M dichromate for titration. What was the percentage of iron in the ore?

SELECTED REFERENCES

W. J. Blaedel and V. W. Meloche, *Elementary Quantitative Analysis,* 2nd ed., Harper and Row, New York, 1963, pp 472, 478, 835. Discussion of basic reactions and sources of error in oxidation-reduction titrations.

H. A. Laitinen and W. E. Harris, *Chemical Analysis,* 2nd ed., McGraw-Hill, New York, 1974, Chapter 17. Detailed discussion of the mechanism of the reaction between dichromate and iron(II).

D. G. Peters, J. Hayes, and G. Hieftje, *Chemical Separations and Measurements,* W. B. Saunders Company, Philadelphia, 1974, Chapter 10.

4–3 PERMANGANIMETRIC DETERMINATION OF IRON IN AN ORE

Background

Potassium permanganate has a venerable analytical history, having been used to titrate iron(II) as long ago as 1846. Although it has the advantage of serving as its own indicator, it also has several disadvantages. One is that it is not a primary standard. Another is that it slowly reacts with traces of organic matter always present in distilled water to produce manganese dioxide, which catalyzes the decomposition of permanganate. After preparation, therefore, permanganate solutions should stand for at least a day and then be filtered to remove manganese dioxide before being standardized.

Sodium oxalate is a common primary standard for permanganate solutions. The mechanism of the reaction between permanganate and oxalate is complex, and accurate results are obtainable only under controlled conditions. Under optimum conditions positive and negative errors still occur, but effectively cancel each other. Divalent manganese and higher temperatures increase the rate of reaction, which otherwise is exceedingly slow. At high temperatures, however, the decomposition of both permanganate and oxalate in solution introduces errors. Also, side reactions such as formation of hydrogen peroxide may occur; if hydrogen peroxide is formed, care must be taken that it remains in solution to react with more permanganate and is not lost by decomposition to water and oxygen. Decomposition is accelerated by vigorous stirring, so mixing during oxalate titrations must be gentle. The most reproducible results in the standardization titrations are obtained by adding about 90% of the required amount of permanganate to an oxalate–sulfuric acid solution at room temperature. This mixture is allowed to stand until the permanganate color disappears; then the solution is heated to about 55°C and the titration finished slowly.

Permanganimetric Oxidation of Iron

In this experiment a sample of iron ore is analyzed for iron by titration with a solution of potassium permanganate. The permanganate is made up approximately and then standardized against sodium oxalate.

The dissolution of iron ores, and subsequent reduction of iron to the divalent oxidation state with tin(II) are discussed in the background section of Experiment 4-2. Two minor modifications are introduced. One is that the reduced iron solution is diluted prior to titration so that oxidation of chloride by permanganate is minimized. The other is that manganese(II) sulfate is added along with phosphoric and sulfuric acids to accelerate the otherwise slow iron(II)–permanganate reaction. A mixture of manganese(II) sulfate, phosphoric acid, and sulfuric acid often is called Zimmermann-Reinhardt reagent.

The reaction for the titration of iron with permanganate is

$$5Fe^{++} + MnO_4^- + 8H^+ \rightarrow 5Fe^{3+} + Mn^{++} + 4H_2O \qquad (4-5)$$

The important reaction in the standardization of the permanganate titrant is

$$2MnO_4^- + 5H_2C_2O_4 + 6H^+ \rightarrow 2Mn^{++} + 10CO_2 + 8H_2O \qquad (4-6)$$

The use of Zimmermann-Reinhardt reagent introduces several additional reactions involving transient intermediate oxidation states of manganese. The reaction mechanism, like that of the dichromate-chromium(III) system, is not completely understood. (See reference by Laitinen and Harris at end of section for detailed discussion.) A brief discussion of prereductants for iron is included in Experiment 4-2.

Procedure (median time 5.5 hr)

Preparation and Standardization of 0.02 M KMnO$_4$ Solution

Weigh about 3.0 g of KMnO$_4$ on a triple-beam balance and dissolve in about 1 liter of distilled water. Store the solution at least a day to allow the permanganate to oxidize organic matter present in the solution. Filter the aged solution through a large sintered-glass filter, using a suction-flask assembly.[1] Store in a bottle previously rinsed with a small portion of the filtered solution.

---

[1] The glass filter can be cleaned by rinsing, first with HCl and then several times with water.

Dry 1.0 to 1.5 g of $Na_2C_2O_4$ for 1 to 2 hr at $110°$ C. Weigh, to the nearest 0.1 mg, samples of 0.22 to 0.28 g into 200-ml conical flasks. Dissolve each sample in 50 ml of water and add 30 ml of 3 M H_2SO_4. Calculate the approximate amount of $KMnO_4$ solution needed to react with a sample of sodium oxalate; it should be in the range from 35 to 45 ml. With stirring add about 90% of the calculated amount to the cold oxalate-sulfuric acid solution. Allow to stand at room temperature until it turns colorless, a process that should take no more than 5 min. Warm the solution to 50 to $60°C$. (At this temperature the flask will feel hot but not uncomfortable to hold.) Titrate slowly to a pink color. Calculate the molarity of the permanganate.

Dissolution and Prereduction of Sample

Weigh, dissolve, and prereduce a set of iron ore samples by the procedure described in Experiment 4-2 under "Procedure for Dissolution of Sample."

Titration of Iron(II) with Permanganate

Dilute the solution that has been reduced with tin(II) to 250 ml, add 25 ml of Zimmermann-Reinhardt solution, and titrate immediately with standard $KMnO_4$ to a faint pink.

Determine a blank by placing 10 ml of 6 M HCl and 2 drops of $SnCl_2$ solution in a 500-ml flask. Add 10 ml of saturated $HgCl_2$ solution, dilute to 250 ml, add 25 ml of Zimmermann-Reinhardt solution, and titrate as for the sample.

Report the percentage of iron in the sample.

Calculations

Calculate the molarity of the permanganate. The following may be used as a guide. At the equivalence point in oxalate titrations

$$\text{moles } KMnO_4 = \text{moles } Na_2C_2O_4\,(2/5)$$

The term (2/5) is derived from Equation (4-6), where 1 mole of permanganate reacts with (5/2) moles of oxalate:

$$\text{moles } Na_2C_2O_4 = \frac{\text{wt } Na_2C_2O_4 \text{ taken}}{\text{wt } Na_2C_2O_4 /\text{mole}}$$

The overall expression for molarity, then, is

$$\text{molarity KMnO}_4 = \frac{\text{moles}\cdot\text{KMnO}_4}{\text{ml KMnO}_4/1000}$$

$$= \frac{(\text{wt Na}_2\text{C}_2\text{O}_4)(2)}{(\text{wt Na}_2\text{C}_2\text{O}_4/\text{mole})(5)(\text{ml KMnO}_4/1000)}$$

At the equivalence point in iron(II) titrations, 5 moles of iron(II) is equivalent to 1 mole of permanganate.

PROBLEMS

4–4.   The iron in a 0.3412-g sample of iron ore, determined by the procedure of this experiment, required 38.62 ml of potassium permanganate solution. A 0.2611-g sample of sodium oxalate required 41.68 ml of the same titrant. What was the percentage of iron in the sample?

4–5.   Titanium(III) is oxidized to titanium(IV) by permanganate. What is the percentage of titanium in a 2.468-g sample that requires 43.96 ml of 0.02317 M permanganate for titration?

4–6.   A 0.5359-g sample of iron ore was dissolved, reduced, and titrated with 43.01 ml of a standard solution containing 2.054 g of pure potassium permanganate in 500.0 ml. What was the percentage of iron in the sample?

4–7.   Complete and balance the following equations:

$\text{BrO}_3^- + \text{AsO}_3^{\,3-} + \text{H}^+ \rightarrow$

$\text{MnO}_2 + \text{Fe}^{++} + \text{H}^+ \rightarrow$

$\text{AsO}_3^{\,3-} + \text{MnO}_4^- + \text{H}^+ \rightarrow$

$\text{FeSO}_4\cdot(\text{NH}_4)_2\text{SO}_4 + \text{NaVO}_3 + \text{H}_2\text{SO}_4 \rightarrow \text{VOSO}_4 +$

$\text{Mn}^{++} + \text{MnO}_4^- + \text{H}_2\text{O} \rightarrow \text{MnO}_2 +$

$\text{Cr}_2\text{O}_7^= + \text{I}^- + \text{H}^+ \rightarrow$

$\text{IO}_3^- + \text{I}^- + \text{H}^+ \rightarrow$

4–8.†   By mistake an analyst standardized a potassium permanganate solution with sodium oxalate, thinking it was oxalic acid dihydrate. On the basis of this standardization the iron content of an ore was reported to be 57.45% iron(III) oxide. What was the correct percentage of iron(III) oxide in the sample?

4–9.† If 20 ml of 0.1 M tin(II) chloride solution is added to 20 ml of 0.1 M iron(III) chloride solution, what will be the iron(III) concentration after reaction is complete?

SELECTED REFERENCES

G. F. Smith, *The Wet Chemical Oxidation of Organic Compositions*, G. F. Smith Chemical Co., Columbus, Ohio, 1965. A discussion of the analytical use of perchloric acid as an oxidant.

F. Margueritte, *Compt. Rend.* **22**, 587 (1846). First use of permanganate for determination of iron.

I. M. Kolthoff, E. B. Sandell, E. J. Meehan, and S. Bruckenstein, *Quantitative Chemical Analysis*, 4th ed., Macmillan, New York, 1969, pp 816 and 828. Good general discussion.

H. A. Laitinen and W. E. Harris, *Chemical Analysis*, 2nd ed., McGraw-Hill, New York, 1974, Chapter 17. A comprehensive discussion of the errors and side reactions involved in permanganate titrations of oxalate and iron(II).

D. G. Peters, J. Hayes, and G. Hieftje, *Chemical Separations and Measurements*, W. B. Saunders Company, Philadelphia, 1974, Chapter 10.

4–4 IODOMETRIC DETERMINATION OF COPPER IN BRASS

Background

The next two experiments illustrate analytical methods involving the iodine–iodide couple:

$$I_2 + 2e^- \rightleftarrows 2I^-   E^0 = 0.53 \text{ V} \tag{4–7}$$

This couple is important because it has a standard electrode potential that permits the analytical use of iodine as an oxidant for substances of lower electrode potential and iodide as a reductant for substances of higher potential. Because its electrode potential is little affected by either pH change or complexing reagents, this couple can be used in conjunction with half-reactions that change potential with pH or with the addition of auxiliary reagents.

Analytically useful applications include those in which solutions of iodine are used to titrate reduced materials directly and those in which oxidizing agents are determined through oxidation of iodide to iodine. In the latter the iodine formed is titrated with a standard solution of sodium thiosulfate, $Na_2 S_2 O_3$.

Iodometric Determination of Copper

In this experiment the copper in a brass sample is determined by a method involving liberation of iodine. The sample is dissolved in nitric acid, and the solution boiled to remove most of the nitrogen oxides formed during the metal oxidation. The residual nitrogen oxides are eliminated by the addition of urea. Complete removal of nitrogen oxides is necessary to prevent iodide oxidation.

Iron, present in most brasses, also causes iodide oxidation. This interference is eliminated by the addition of fluoride, which forms a stable complex with iron(III). Next the pH is adjusted to 3.5 to 4.5, an excess of potassium iodide is added, and the iodine formed is titrated with sodium thiosulfate.

Although the titration of iodine is the only titration of significance for which thiosulfate is used as a standard solution, it is an important one. Other applications of thiosulfate are few. One reason is that oxidizing agents stronger than iodine oxidize it to a mixture of higher oxidation states of sulfur; another is that ions of transition metals such as copper decrease the stability of sodium thiosulfate solutions by catalytic air oxidation. Acids also make thiosulfate solutions unstable by promoting disproportionation to sulfite and elemental sulfur:

$$HS_2O_3^- \rightarrow HSO_3^- + S \qquad (4-8)$$

Addition of a small amount of sodium carbonate to make the thiosulfate solution alkaline prevents this decomposition. Water used for the preparation of standard thiosulfate solutions may be boiled to destroy sulfur bacteria, which find these solutions an attractive medium for growth. Alternatively, a bactericidal agent such as a mercury(II) salt or chloroform may be added.

Any of several primary standards, including iodine, potassium dichromate, potassium iodate, and electrolytic copper metal, may be used to standardize thiosulfate solutions. For this experiment copper is chosen because it is the material being determined and is readily available in primary-standard quality as electrical wire.

The principal reactions involved in iodometric copper analysis are (1) dissolution of sample in dilute nitric acid,

$$3Cu + 8HNO_3 \rightarrow 3Cu^{++} + 2NO + 4H_2O + 6NO_3^- \qquad (4-9)$$

or

$$Cu + 4HNO_3 \rightarrow Cu^{++} + 2NO_2 + 2H_2O + 2NO_3^- \qquad (4-10)$$

(2) removal of residual nitrogen oxides by addition of urea,

$$2NO + O_2 \rightleftarrows 2NO_2 \tag{4-11}$$

$$2NO_2 + H_2O \rightleftarrows HNO_2 + HNO_3 \tag{4-12}$$

and

$$2HNO_2 + O=C\begin{smallmatrix}/NH_2\\\\\backslash NH_2\end{smallmatrix} \rightarrow 2N_2 + CO_2 + 3H_2O \tag{4-13}$$

(3) neutralization of remaining nitric acid with sodium hydroxide, followed by pH adjustment; (4) complexation of iron(III) with fluoride,

$$Fe^{3+} + F^- \rightleftarrows FeF^{++} \tag{4-14}$$

(5) addition of excess potassium iodide,

$$2Cu^{++} + 4I^- \rightleftarrows Cu_2I_2 \text{ (5)} + I_2 \tag{4-15}$$

and (6) titration of iodine with thiosulfate,

$$I_2 + 2S_2O_3^{=} \rightleftarrows 2I^- + S_4O_6^{=} \tag{4-16}$$

Precautions must be taken to avoid side reactions. For instance, iodine will slowly oxidize tetrathionate to sulfate, especially at high pH values. At low pH values sulfurous acid may be formed from thiosulfate [Equation (4-8)]. Another source of error in strong acid solution is air oxidation of iodide:

$$O_2 + 4H^+ + 4I^- \rightleftarrows 2I_2 + 2H_2O \tag{4-17}$$

This is called *oxygen error*. These side reactions may be avoided by carrying out the thiosulfate–iodine titration in the pH range 2 to 5.

The volatility of iodine creates problems. Iodine loss can be minimized by keeping the temperature low, titrating promptly, and adding excess iodide to stabilize the iodine in solution as triiodide:

$$I_2 + I^- \rightleftarrows I_3^- \quad K_{eq} = 7.1 \times 10^2 \tag{4-18}$$

Starch gives an intense blue color with triiodide that serves as a specific indicator. The triiodide ion appears to be just the right size to enter the helical structure of starch and thereby form a colored complex. The color of free iodine, red-brown in high and yellow in low concentrations, also can be used as an indication of its presence, but is not so easily seen as that of the complex with starch. In acidic solutions starch tends to undergo decomposition that is accelerated by high concentrations of iodine. Therefore, addition of starch is best delayed until near the equivalence point of a titration.

Most strong oxidizing agents can be determined iodometrically. An excess of iodide is added, and iodine is produced in an amount equivalent to that of the oxidant present in the sample. The liberated iodine then is titrated with thiosulfate.

Procedure (median time 3.9 hr)

Preparation of Starch Indicator Solution

Mix 1 g of soluble starch with enough cool water to produce a thin paste.[1] (Dry starch does not dissolve readily in boiling water.) Add this suspension to 100 ml of boiling water and boil the solution for about 2 min. Solutions of starch begin to decompose and give poor end points after a day or two, so discard them after that time.

Standardization of 0.1 M $Na_2S_2O_3$ Solution

Dissolve 25 g of $Na_2S_2O_3 \cdot 5H_2O$ and 0.1 g of Na_2CO_3 in a liter of distilled water. Add a drop of chloroform.

Weigh accurately 0.2-g samples of clean copper wire into 200-ml conical flasks.[2] Add 10 ml of 6 M HNO_3 to each in a fume hood. When dissolution is complete, boil the solution for a short time to remove most of the nitrogen oxides. Add 10 ml of water and 5 ml of 4% urea solution, and boil again for about 1 min. Cool. When ready to titrate, add about 30 ml of water and then add 2.5 M NaOH until a slight permanent precipitate of $Cu(OH)_2$ is obtained. This will require about 15 to 20 ml of NaOH, depending on the amount of HNO_3 present. Add 1 to 2 g of ammonium acid fluoride, NH_4HF_2, and swirl until dissolved.

Cool, add 3 g of KI, and titrate immediately to near the end point

---

[1] Commercial preparations, stable and usable as obtained, are available from chemical suppliers. These work well and eliminate the need for frequent preparation of starch solutions.

[2] Trace impurities markedly increase the resistance of copper wire. For electrical use they are removed by electrolytic refining to a level well below a part per thousand. Copper produced for electrical wiring is therefore an excellent primary standard. The thin coat of oxide sometimes present on the wire can be removed by polishing with fine emery cloth, followed by wiping with clean toweling.

with $Na_2S_2O_3$. When the solution has become pale yellow or buff, add 5 ml of fresh starch solution and titrate to the first complete disappearance of blue.[3]

Procedure for Sample

Weigh 0.2-g samples of brass into 200-ml conical flasks. Dissolve in HNO_3.[4] Add urea and boil as described for the standards. When ready to titrate, neutralize with sodium hydroxide and dilute to 50 ml with water. Add 1 g of ammonium acid fluoride and swirl until dissolved. Add 3 g of KI and complete the determination as in the standardization. Remember to use alternation of standards and samples.

Calculate the molarity of the $Na_2S_2O_3$ solution. One mole of copper requires 1 mole of thiosulfate for titration. Calculate and report the percentage of copper in the sample.

PROBLEMS

4–10.   A 0.4121-g brass sample was dissolved, and after addition of excess iodide the copper present was titrated with 0.1073 M thiosulfate. If 38.40 ml of thiosulfate was required to the starch end point, what was the percentage of copper in the sample?

4–11.   A 1.620-g sample of copper ore, analyzed for copper by the method of this experiment, required 43.50 ml of 0.1080 M thiosulfate for titration. What was the copper content of the sample?

4–12.   The molar solubility of iodine in water and dilute salt solutions is 1.34×10^{-3} M. Using the information in Equation (4-18), calculate the total solubility of iodine ($[I_2]$ plus $[I_3^-]$) in 0.100 M potassium iodide.

4–13.†   In a solution 0.01 M in KI, what I_2 concentration is required to give a blue color with starch? Assume the color can be seen when the triiodide concentration reaches 10^{-5} M.

4–14.   Potassium iodate is sometimes used as an oxidizing titrant in

---

[3] To avoid etching of flasks by HF, empty and rinse immediately after completing each titration.
[4] If the brass sample contains tin, a white precipitate of metastannic acid, $SnO_2 \cdot xH_2O$, will remain; it will not affect the titration and can be ignored.

place of iodine. With acid and excess iodide, iodine is formed quantitatively according to the reaction

$$IO_3^- + 5I^- + 6H^+ \rightleftarrows 3I_2 + 3H_2O$$

If a 1.5447-g sample containing arsenic(III) requires 31.89 ml of 0.1121 M potassium iodate solution for titration of the arsenic(III) to arsenic(V), what is the percentage of arsenic in the sample?

4–15.   A solution of sodium thiosulfate was standardized by titration of the iodine formed when excess acid and potassium iodide were added to a 0.2500-g sample of pure potassium dichromate. What was the molarity of the thiosulfate if 41.02 ml was required?

4–16.† Calculate the equilibrium constant for the reaction

$$Sn^{4+} + 2I^- \rightleftarrows Sn^{++} + I_2$$

4–17.   Which of the following pairs contains substances that would be expected to react with each other to an appreciable extent:
(a) AsO_3^{3-} and $Fe(CN)_6^{4-}$; (b) AsO_3^{3-} and $Fe(CN)_6^{3-}$; (c) AsO_4^{3-} and $Fe(CN)_6^{4-}$; (d) AsO_4^{3-} and $Fe(CN)_6^{3-}$?

SELECTED REFERENCES

I. M. Kolthoff, E. B. Sandell, E. J. Meehan, and S. Bruckenstein, *Quantitative Chemical Analysis,* 4th ed., Macmillan, New York, 1969, p. 854.

H. A. Laitinen and W. E. Harris, *Chemical Analysis,* 2nd ed., McGraw-Hill, New York, 1974, Chapter 19.

D. G. Peters, J. Hayes, and G. Hieftje, *Chemical Separations and Measurements,* W. B. Saunders Company, Philadelphia, 1974, Chapter 10.

4–5   ORGANIC FUNCTIONAL-GROUP ANALYSIS: DETERMINATION OF ETHYLENE GLYCOL BY PERIODATE CLEAVAGE

> I believe that our souls are in our hands, for we do everything to the world with our hands. Sometimes I think we don't use our hands half enough; it's certain we don't use our heads.
>
> *Ray Bradbury*

Background

In this experiment an organic compound, ethylene glycol, is determined by oxidative cleavage with potassium periodate. Although the same fundamental analytical techniques are applicable to both organic and inorganic compounds, inorganic samples are used more frequently in introductory analytical courses because they tend to be less sensitive to experimental variables. One difficulty associated with organic analysis is that many compounds can undergo several reactions simultaneously; another is the number and variety of structures and combinations that possess similar chemical properties. Accurate, reproducible analyses therefore usually require practice and careful technique.

As implied above, the determination of organic compounds is in some respects as much an art as it is a science. In the past, brilliant scientific artists often outdistanced available fundamental knowledge to solve problems by intuition and exceptional technique. Today the scientific basis for methods of organic analysis is constantly being explored and expanded, and existing methods are steadily being improved or replaced by better ones.

Quantitative organic analysis may be conveniently divided into elemental and functional-group determinations. In *elemental analysis* a compound is broken down under drastic conditions to simple measurable species. For example, an organic compound may be analyzed for carbon and hydrogen by heating a few milligrams in a stream of oxygen until the carbon is converted to carbon dioxide and the hydrogen to water. The gas stream is then passed in sequence through a preweighed tube of magnesium perchlorate, which absorbs the water, and one of sodium hydroxide, which absorbs the carbon dioxide. From the increase in weight of the tubes the percentages of carbon and hydrogen in the sample are calculated.

In *functional-group analysis,* atoms or groups of atoms having specific reactivity are measured. Although in a broad sense any two atoms covalently bonded can be considered a functional group, the term is generally applied to groups of atoms that undergo specific reactions, such

as aldehydes ($-\overset{\overset{\text{O}}{\|}}{\text{C}}-\text{H}$), alkenes ($-\text{C}=\text{C}-$), alcohols ($-\text{OH}$), and nitro groups ($-\text{NO}_2$). Both qualitative and quantitative determinations of functional groups are important; such information helps determine structure, reactivity, and purity. For example, compounds that have hydroxyl groups on adjacent carbon atoms are oxidized selectively by several reagents, among them periodate. Ethylene glycol is the simplest compound of this type.

Owing to improved speed and convenience, applications of chemical methods of functional-group analysis have increased steadily in recent years. The development of better and faster procedures is demonstrated by the number of books devoted solely to this subject.

Functional-group determinations can be carried out by physical as well as chemical techniques. Physical methods, especially those using instruments, have made possible many kinds of measurements never before envisioned. A modern chemist would be severely handicapped without access to the information provided by techniques such as spectrophotometry, chromatography, and electrochemistry. Most physical methods of analysis can be classed as either spectral or electrochemical.

Oxidation of Ethylene Glycol

This experiment illustrates functional-group analysis by a chemical method. Ethylene glycol reacts with periodate (added as periodic acid) in a 1:1 mole ratio to yield formaldehyde and iodate:

$$\begin{array}{c} H_2COH \\ | \\ H_2COH \end{array} + IO_4^- \rightleftharpoons H_2O + \begin{array}{c} H_2C-O \\ | \quad\quad IO_3 \\ H_2C-O \end{array} \rightarrow 2H_2CO + IO_3^- \qquad (4{-}19)$$

Under optimum conditions this reaction goes to completion, while side reactions are negligible ($<0.1\%$).

The rate of glycol reaction, as with many organic reactions, is slow but increases with periodate concentration. Thus it can be driven to completion in a reasonable time by an excess of periodate. If the total amount of periodate added is known, and if the excess is treated in turn with a known excess of standard arsenite, titration of the remaining arsenite with standard iodine solution will give the amount of periodate left after all the glycol has reacted. The reactions are

$$IO_4^- + H_2AsO_3^- \left(\begin{array}{c} \text{known} \\ \text{excess} \end{array}\right) \rightarrow IO_3^- + H_2AsO_4^- \qquad (4{-}20)$$

and

$$H_2AsO_3^- + I_2 + H_2O \rightleftarrows H_2AsO_4^- + 2I^- + 2H^+ \qquad (4{-}21)$$

The initial reaction of periodate with glycol to form a 5-membered ring is reversible [double arrows in Equation (4-19)]. Equilibrium is attained rapidly because the rates of the forward and back reactions are faster than the irreversible reaction that follows [single arrow in Equation (4-20)]. The reversible step is not of direct analytical use because the equilibrium constant is only about 18 at pH 1 and about 1000 at pH 8. Although the equilibrium constant is larger at high pH values, even a constant on the order of 1000 would seldom be analytically suitable, and the reaction at pH 1 becomes useful only because the second part of the reaction is irreversible. As the complex breaks down, more of the intermediate is formed until all the glycol is oxidized to formaldehyde.

The rate of decomposition of the 5-membered periodate-ring intermediate is the important factor in timing the experimental operations. At pH 1 and room temperature the time required for half the intermediate to decompose (the half-time) is about 1 min. At pH 8 it is several hours. Thus the kinetics of the reaction are such that for completion of the experiment within a reasonable time a low pH is necessary even though the equilibrium constant for the initial, reversible step of the reaction is not favored in acid solution. In 10 half-times (about 10 min at pH 1) the overall reaction is quantitative, that is, 99.9% complete. In the overall process, then, a slow, irreversible second step outweighs the effect of an unfavorable equilibrium constant for the first step.

Two side reactions may affect the analytical results. Periodate may react further with formaldehyde, or it may decompose spontaneously. If the rate of a side reaction is 1000 times or more smaller than that of the desired reaction, it should not cause significant error in the analytical procedure. In this analysis enough time is allowed for the main reaction to go to completion (10 to 15 min). The reaction is next stopped, or quenched, by raising the pH with bicarbonate and then adding excess arsenite to react with the remaining periodate. Since the spontaneous decomposition (possibly photodecomposition) of periodate proceeds rapidly in bicarbonate solutions, arsenite must be added without delay once the pH has been raised. Because of this decomposition reaction the excess periodate cannot be titrated directly with arsenite, and so an indirect procedure must be used—addition of excess arsenite followed by titration with iodine.

This experiment illustrates the use of the method of back titration of excess standard reagent, a convenient technique for slow reactions. Many precise volumetric measurements are required; carelessness in any one will invalidate the results.

All experimental work other than the preparation of some of the solutions must be completed in a single session. The technique of alternation is recommended for the preparation and quenching steps as well as for the titrations. Because of the time required for the numerous volumetric measurements, it is suggested that this analysis be performed only in triplicate. Careful planning is essential.

Procedure (median time 4.4 hr)

Preparation of Standard 0.14 M Arsenite Solution from As_2O_3 (As_4O_6)

Weigh to the nearest 0.1 mg about 3.6 g of dry primary-standard As_2O_3 into a dry beaker and add 50 ml of 2.5 M NaOH.[1] When the As_2O_3 is dissolved, transfer the solution quantitatively to a 250-ml volumetric flask and half fill the flask with water. Add 15 ml of 6 M HCl, swirl, and then add 8 g of $NaHCO_3$.[2] Dilute to volume. (*Caution:* Arsenic compounds are poisonous.)

Preparation of 0.06 M Iodine Solution

Weigh about 8.4 g of I_2 and 17 g of KI into a 1-liter storage bottle.[3] Add about 15 ml of water and allow at least 30 min for the I_2 to dissolve, shaking occasionally.[4] Dilute to 500 ml with water. Prepare starch indicator solution as described in Experiment 4-4.

Preparation of Periodic Acid Solution

Dissolve about 5.2 g of H_5IO_6 in 100 ml of water.[5] Prepare on the same day it is to be used.

Analysis

Obtain a glycol sample. Dilute to volume in a 100-ml volumetric flask with distilled water and mix.

---

[1] As_2O_3 is only slowly soluble in neutral or acidic solutions, so it is dissolved first in base. About 10 min is required.

$$As_2O_3 + 4OH^- \rightarrow 2HAsO_3^= + H_2O$$

The molecular unit for arsenic(III) oxide is As_4O_6; this formula is sometimes recommended in preference to As_2O_3. However, the latter is in widespread use and causes no difficulty in analytical work where the first step is dissolution of the compound. Arsenic(III) in aqueous solution exists in several forms, including H_3AsO_3. We use here H_3AsO_3.

[2] $NaHCO_3$ is added to adjust the solution to pH 7 to 9. If too much or too little acid is added, the amount of $NaHCO_3$ used here will not give the correct pH, and the stability of the arsenite solution will be impaired.

[3] I_2 is both volatile and corrosive. Use a triple-beam balance, not an analytical balance. Always weigh I_2 on a watch glass rather than a weighing paper, and clean up any spillage immediately.

[4] The solubility of I_2 in water is only about 0.3 g/liter, but is greatly increased if iodide is present to form triiodide ion, I_3^-. Even with iodide present, the rate of dissolution is slow, and if the iodide concentration is reduced by dilution, the rate decreases still further. The intense color of iodine makes it difficult to tell when the last particles have dissolved.

[5] Periodic acid is a strong acid and the anion is a strong oxidant. Therefore, periodic acid and periodates, like iodine, should not be weighed on an analytical balance or in contact with reactive materials such as paper.

1.   *Preparation*
a.   Blanks. Into one set of flasks pipet 10 ml of H_5IO_6 solution and add about 10 ml of distilled water.
b.   Standards. To a second set of flasks add about 20 ml of water.
c.   Samples. Into a third set of flasks containing 10-ml aliquots of the glycol sample, pipet 10 ml of periodate solution. Swirl gently to mix, and allow to stand for 10 to 15 min.
2.   *Quenching.* After 10 to 15 min quench the blanks, standards, and samples. Add 3 g of $NaHCO_3$, swirl to partially dissolve the $NaHCO_3$, and at once pipet 20 ml of standard Na_3AsO_3 into each flask.[6]
3.   *Titration.* Add 5 ml of starch indicator, and titrate with I_2 solution to the first blue color that persists after swirling for 10 to 20 sec.[7,8]

Report the weight of ethylene glycol in the entire sample.

Calculations

The molarity of the I_2 solution is calculated from the standardization titrations and should be about 0.06 to 0.07 M. The weight of ethylene glycol in grams is calculated from the expression

$$\text{wt glycol} = \frac{(V_{\text{sample}} - V_{\text{blank}})(M_{I_2})(\text{mol wt glycol})(100)}{(1000)(V_{10})}$$

(4–22)

where V_{sample} is the milliliters of I_2 solution required to titrate the sample, V_{blank} is the milliliters of I_2 solution required to titrate the blank,[9] and V_{10} is the calibrated volume of the pipet. In the calculation use values of sample 1, blank 1, and standard 1 together; then sample 2, blank 2, and standard 2; and so forth. If one of the titration results is rejected on the basis of the Q test, substitute for it the value obtained nearest in time.

---

    [6] The timing here is critical. Once the $NaHCO_3$ has been added, periodate begins to decompose rapidly. Add the arsenite solution as quickly as possible after the bicarbonate has been mixed with the flask contents. After the addition of arsenite, the solutions are relatively stable and can stand several hours before titration if necessary.
    [7] I_2 is somewhat volatile, even from aqueous solution. Errors due to volatility can be minimized by placing an inverted test tube on the buret and by using alternation.
    [8] The titration volumes should be in the following ranges: blank, 5 to 10 ml; standard, 40 to 45 ml; sample, 30 to 35 ml.
    [9] Note that the blank in this analysis is not the same as an indicator blank in conventional direct titrations. Therefore V_{blank} should not be subtracted from the volume of titrant used in the standardizations.

PROBLEMS

4–18.   An arsenite solution was prepared by dissolving 2.473 g of arsenic trioxide and diluting to 250.0 ml. A 25.00-ml portion of this solution required 40.17 ml of iodine solution. Calculate the molarity of the iodine solution.

4–19.   In the analysis of a sample of ethylene glycol by the procedure of this experiment, the following data were obtained: blank titration, 11.76 ml; standard titration, 46.26 ml; sample titration, 32.69 ml. If the weight of As_2O_3 taken was 3.6019 g and the calibrated volume of the 10-ml pipet was 9.988 ml, what was the weight of glycol in the sample?

4–20.†   Is it theoretically feasible to titrate a solution of H_3AsO_4 with a solution of iron(II) sulfate in a buffer of pH 3? Explain.

4–21.†   Calculate the potential of a platinum electrode in a solution that is 0.5 M in H_3AsO_3, 0.1 M in H_3AsO_4, and 0.2 M in HCl.

SELECTED REFERENCES

N. D. Cheronis and T. S. Ma, *Organic Functional Group Analysis,* Wiley (Interscience), New York, 1964.

F. E. Critchfield, *Organic Functional Group Analysis,* Pergamon Press, Elmsford, N.Y., 1962.

R. D. Guthrie in *Methods of Carbohydrate Chemistry, Vol. I, Analysis and Properties of Sugars,* R. L. Whistler and M. L. Wolfrom, Eds., Pergamon Press, Elmsford, N.Y., 1962, p 432.

G. H. Schenk, *J. Chem. Educ.* **39**, 32 (1962).

S. Siggia, *Quantitative Organic Analysis via Functional Groups,* 3rd ed., Wiley, New York, 1963.

D. G. Peters, J. Hayes, and G. Hieftje, *Chemical Separations and Measurements,* W. B. Saunders Company, Philadelphia, 1974, Chapter 10.

Chapter 5

GRAVIMETRIC ANALYSIS

As a beginning chemist, ... make a habit of weighing carefully, of pouring from one vessel to another without spilling and without missing the last drop, and of observing the small details which if overlooked often spoil several weeks of careful work.

J. J. Bertzelius

5-1 GENERAL BACKGROUND

Gravimetric analysis at one time constituted a major part of quantitative analysis, but today occupies a less prominent place. Therefore, instruction on this topic is minimized in keeping with its lesser importance. The trend is to avoid gravimetric analysis in routine work because of the time and technique required, although one time-consuming step, weighing, has become less burdensome with the introduction of single-pan balances. In general, gravimetric methods take longer than volumetric ones because more operations are required. Volumetric analysis generally involves the steps of sampling, drying, weighing, dissolving, adjusting conditions, and finally titrating. Gravimetric analysis involves all but titrating, and also precipitating, digesting, filtering, washing, igniting, and another weighing. Gravimetric methods have the advantage of broad applicability (Figure 5-1) and may be used in the same range and give about the same precision as volumetric methods.

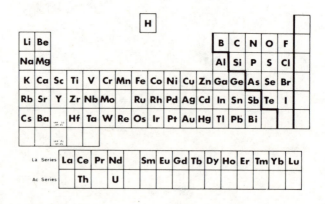

FIGURE 5-1. Elements determinable by gravimetric methods.

Coprecipitation, however, is more serious than in volumetric precipitation titrations because any nonvolatile contaminant causes error. Despite the disadvantages, a gravimetric procedure may be preferred when only a few analyses are needed and when the preparation and standardization of volumetric reagents is time consuming.

Precipitation

Precipitation occurs in two steps—nucleation and crystal growth. *Nucleation* is the process of formation within a supersaturated solution of the smallest particles capable of growth to larger particles. *Crystal growth* consists of deposition of material, probably by adsorption, from solution onto nuclei and particles already formed.

The importance of nucleation lies in the fact that the average final particle size of the precipitate is determined by the number of nuclei initially formed. This number, in turn, is determined by the number and effectiveness of the nucleation sites and the extent of supersaturation. The extent of supersaturation is the variable most amenable to experimental control.

Coprecipitation

Coprecipitation is contamination of a precipitate by normally soluble materials. The main types are adsorption and occlusion. *Adsorption* is the attraction of solute particles to a surface by the action of electrostatic and van der Waals forces. For ionic substances, electrical attraction to localized centers of residual positive or negative charge is the most important factor.

The extent and kind of ionic adsorption can be predicted qualitatively on the basis of solubility, sign and magnitude of charge on the ion, ion size, and concentration. Because adsorption is a surface phenomenon, its effect can be reduced by keeping the surface area small.

Occlusion is internal contamination of a precipitate by normally soluble substances. As individual small crystals grow, foreign adsorbed material occupying lattice positions on the crystal may be surrounded. This type of coprecipitation is more important than adsorption because it affects even precipitates with small surface area.

The extent and kind of contamination caused by occlusion can be predicted on the same basis as that of adsorption. The history of the precipitate also is involved—the order of reagent addition affects the kind of occlusion taking place. Occlusion of foreign anions is minimized if an excess of the anion being precipitated is present during crystal growth; similarly, occlusion of foreign cations is minimized if an excess of the cation being precipitated is present. Thus the historical factor can be explained in terms of adsorption during crystal growth.

Important considerations in the experimental control of occlusion are concentration of impurities, temperature during precipitation, charge on foreign ions, order of mixing, speed of precipitation, and aging after precipitation.

An effective way to reduce coprecipitation is to keep the supersaturation ratio low during precipitate formation. *Supersaturation ratio* refers to the extent to which the solubility of a precipitate is exceeded during the addition of the precipitating reagent. A low ratio results in larger particles of precipitate and so makes filtration and washing easier; it also reduces contamination by foreign ions and thereby minimizes occlusion and adsorption effects. The supersaturation ratio can be kept low by use of hot, dilute solutions for the precipitation and by addition of the precipitant slowly with efficient stirring.

Aging, or digesting, improves purity and ease of filtration because small particles of a material are more soluble than large particles. A solution in contact with a variety of particle sizes will become supersaturated with respect to the larger particles, saturated with respect to the intermediate ones, and unsaturated with respect to the smallest. The small ones dissolve and liberate occluded material, whereas the large ones grow larger and purer. Because solubilities generally increase with temperature, so does the rate of aging.

Techniques in Gravimetric Analysis

After a test for complete precipitation, the precipitate is separated from the mother liquor by filtration and washed free of soluble salts. Types of filters available include paper, sintered glass, porous porcelain, and glass or asbestos fiber mats in Gooch crucibles. Gooch crucibles are made of porcelain and have a perforated bottom upon which a glass or asbestos fiber mat is laid. Glass fiber mats are available in ready-to-use form. They are more convenient than asbestos as long as the precipitate is being ignited at moderate temperatures, but glass fibers fuse readily in the direct heat of a Bunsen burner. Sintered-glass or porcelain crucibles with

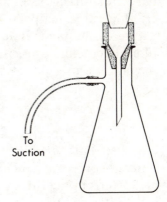

FIGURE 5–2. Cross section of suction filtration assembly with filter crucible.

To
Suction

built-in porous bottoms can be used at higher temperatures than glass fiber mats, although they are difficult to clean. Filtration with paper is discussed below.

For filtrations other than through paper, a suction filtration assembly is used (Figure 5-2). When suction is provided by an aspirator and the filtrate is to be saved, a safety bottle is placed between aspirator and flask to prevent accidental suckback of tap water into the flask. The filtration operation is carried out in the following way. Place the filtering crucible, which has been brought to a known constant weight,[1] in position over a clean suction flask and pour the supernatant liquid through the filter. Leave behind as much as possible of the precipitate so it will not slow the flow through the filter. Use of a stirring rod to guide the solution from the beaker tip to the crucible prevents splashing and keeps solution from running down the outside of the beaker when pouring is interrupted. Next transfer the precipitate to the crucible with a minimum of wash liquid. Use a rubber policeman to loosen precipitate adhering to the beaker walls. After the precipitate has been quantitatively transferred, wash it with successive small portions of wash liquid to remove soluble salts. If the precipitate is sufficiently insoluble, hot wash liquid often can increase the solubility of impurities and increase the filtration rate. In some cases pure water is satisfactory; in others an electrolyte in the water is needed to prevent peptization. Often a solvent mixture will reduce solubility losses. Each portion of wash liquid should be small and be allowed to drain completely before the next is added. If possible, test the completeness of washing to ensure that all the mother liquor has been removed.

After washing, dry the crucible and contents to constant weight. If a

---

[1] The weight of a crucible can be considered constant when it remains the same within 0.2 to 0.3 mg after successive heating and cooling cycles.

burner or furnace is employed, this step is called ignition. Most ignitions are done at a temperature between 200 and 1000°C, depending on the nature of the precipitate. When a burner is used to ignite a precipitate in a filtering crucible, the crucible is placed inside a larger crucible to protect the mat and precipitate from direct contact with burner gases.

Filtration Through Paper

For most purposes, filter paper is not so convenient as a filtering crucible, but for high-temperature ignitions is less expensive than a sintered-porcelain crucible. Also, the problem of removal of ignited material from the filter is avoided.

Filter funnels are available with apical angles of either 58 or 60° Correct fitting of the filter paper to the funnel is important. To prepare a filter, select paper of the proper porosity and diameter and fold it in half, then in quarters (see Figure 5-3). For a funnel of 60° make the quarter fold so that the corners do not coincide but are displaced about 3 mm in each direction. Tear off a small triangular portion of the short corner to make a tight fit between paper and funnel. The second fold is made off center so as to form a cone upon opening whose apical angle is somewhat greater than 60°. A filter folded in this way will fit against the funnel wall near the top, but leave a slight space farther down for better liquid drainage. Open the large quarter of the paper to form a cone. Seat the cone in the funnel, moisten it with water, and press the upper portion gently against the glass to make a seal. When a filter is properly fitted, the stem of the funnel will fill when water is poured into the funnel and remain full after the funnel has drained. Slight suction exerted on the underside of the filter by the weight of this liquid column markedly increases the filtration rate, especially for gelatinous precipitates such as aluminum hydroxide.

Three steps are involved in the actual filtering operation—decanting, washing, and precipitate transfer. *Decanting* is the process of pouring the

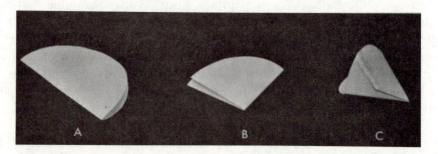

FIGURE 5–3. Method of folding filter paper. *A*, fold along the diameter of the paper. *B*, fold so that the edges do not match. *C*, tear part of corner for closer fit of the upper edge to the funnel.

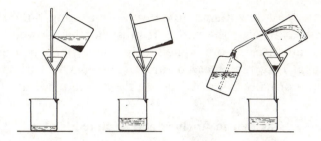

FIGURE 5–4. Operations of decantation and precipitate transfer.

liquid above the precipitate through the filter. Use a stirring rod to guide the solution (Figure 5-4). Next wash the precipitate in the beaker by adding, stirring, and decanting several portions of wash solution. Finally transfer the precipitate to the filter with a stream of wash liquid (Figure 5-4). Use a policeman to dislodge particles of precipitate adhering to the beaker.

After washing is complete, carefully lift out the filter paper with precipitate, fold in the upper edges of the paper, and place in a crucible previously ignited to constant weight. Put the folded side of the paper down in the crucible so that it will not unfold while drying. Collect any small particles of precipitate adhering to the beaker by wiping with a small portion of moistened filter paper. Add it to the crucible. Then place the crucible on a triangle over a burner as illustrated in Figure 5-5. Dry the

FIGURE 5–5. Position of crucible during drying and ignition of a precipitate. [Burner shown is a special type (Meker) sometimes used to attain higher temperatures.]

paper first with a low flame to prevent spattering, and then slightly increase the heat to char the paper. It must not be allowed to burst into flame, which might carry out some precipitate. After smoking has ceased, the burner heat may be increased further until all the carbon is removed. Final ignition normally is carried out at the maximum heat of the burner.

Gravimetric Analysis and Atomic Weights

In the history of chemistry no other problem has undergone the continuous investigation accorded that of defining and measuring atomic weights. Work in this field has included some of the most fascinating chemistry and interesting personalities encountered in science. The problem of atomic weights first revealed itself in 1803 when Dalton proposed that continuous division of any substance would ultimately yield particles that were no longer divisible. For a half century after Dalton's suggestion, enormous confusion existed on the subject because, in the first place, many doubted the existence of atoms. Almost 100 years passed before the last of the doubters disappeared. Part of the reason for this lack of acceptance probably was that Dalton discredited his own ideas by poor experimental work. Another source of confusion was that relative combining weights of different atoms had not yet been clearly established. A third significant difficulty resulted from the use of different reference points by different workers. This problem arose because atomic weights are simply a series of numbers indicating the relative weights of atoms, and some single substance must be chosen as a point of reference. About 1810 Dalton suggested a value of 1 for hydrogen, in 1814 Wollaston suggested oxygen equal to 10, in 1825 Thompson suggested oxygen equal to 1, and in 1830 Berzelius suggested oxygen equal to 100. In 1860 a proposal by J. S. Stas that a value of 16.0000 for oxygen be the standard won general acceptance, and the confusion finally diminished.

Although Dalton and other workers of his time were really determining combining, rather than atomic, weights, they did not clearly recognize the difference between these two concepts, and the picture was not clear for many decades. For this reason the atomic weights they reported were often simple multiples or fractions of the atomic weights in use today.

By 1840 about 100 determinations of atomic weights had been made. Although many others contributed, Dalton (1766-1844) and Berzelius (1779-1848) produced the most important work. Methods of breaking down (analyzing) or building up (synthesizing) selected compounds as methods for atomic weight determinations were developed and refined. Table 5-1 lists some early values.

During the second half of the nineteenth century (1850 to 1890) about 500 determinations of atomic weights were carried out. This work

TABLE 5—1. ATOMIC WEIGHTS AS DETERMINED BY
DALTON AND BERZELIUS
(all values calculated relative to carbon equal to 12)

	Dalton	Berzelius
Hydrogen	1.1	1.007
Carbon	11	12.23
Nitrogen	11	14.16
Sulfur	28	32.19
Silver	114	108.13
Lead	206	207.11

required experimental technique of exceptional accuracy, and only one or two investigators in each generation possessed the combination of intelligence, persistence, and dexterity to report results of a quality that could win general acceptance. One of these was Stas (1813-1891), who spent his life working in this field. His contributions were outstanding; some values he obtained are listed in Table 5-2.

During the next half century (from about 1890 to 1940) about 800 additional determinations of atomic weights were completed. This period virtually marked an end to determinations of atomic weights by chemical means. (The last appears to have been that of fluorine by Scott and Ware in 1957.) The monumental work of T. W. Richards (1868-1928) and his colleagues stands out in this period. Richards took up where Stas left off and, like Stas, used silver as the basis for his work. He found that even the most carefully purified $KClO_3$ always contains some KCl, that precipitated $AgCl$ contains traces of KCl that cannot be removed by washing, and that metallic silver will always contain some oxygen unless stringent precau-

TABLE 5—2. COMPARISON OF SOME 1971 ATOMIC-WEIGHT VALUES
WITH THOSE OBTAINED BY EARLY WORKERS

	Dalton (1810)	Berzelius (1836)	Stas (1890)	Richards and Coworkers (1923)	IUPAC-IUPAP (1971)
Hydrogen	1.1	1.007	1.002	—	1.0079
Carbon	11	12.23	11.99	12.000	12.011
Nitrogen	11	14.16	14.055	14.0080	14.0067
Sulfur	28	32.19	32.069	32.064	32.06
Chlorine	—	35.41	35.456	35.4561	35.453
Silver	114	108.13	107.93	107.876	107.868
Lead	206	207.11	206.905	207.19	207.2

tions are taken. To avoid such problems, Richards had to develop many new techniques for the purification and handling of chemicals. Since he found that water dissolves traces of silica from glass, he substituted platinum or quartz vessels. He developed the centrifuge and also a bottling apparatus in which substances could be heated in any gas or under vacuum under reproducible conditions and subsequently weighed without exposure to air.

Richards' first goal was to revise the atomic weights of silver, chlorine, and nitrogen relative to oxygen. His measurements involved determination of the weight of silver chloride and of silver nitrate produced from a known weight of silver and determination of the weight of silver chloride produced from a known weight of ammonium chloride. With the assumption of oxygen equal to 16.0000 and hydrogen equal to 1.0076 (determined by Morley in 1895), the atomic weights of silver, chlorine, and nitrogen could be calculated. To give an idea of the careful work involved, the procedure used by Richards is outlined here.

Pure silver was obtained from silver nitrate that had been recrystallized fifteen times and fused on lime in an atmosphere of hydrogen. The silver was dissolved in highly purified nitric acid, and silver chloride precipitated by the addition of highly purified sodium chloride. The reactions were carried out in red light to minimize photodecomposition of the precipitate. The amount of silver chloride present in the filtrate was measured with the aid of a device developed by Richards, the nephelometer (a procedure that marked the birth of instrumental analysis). With every possible precaution the silver chloride precipitate then was filtered, washed, dried, and weighed.

Pure ammonium chloride was prepared from ammonium sulfate by treatment with permanganate to oxidize organic matter; ammonia was driven off with pure calcium oxide and absorbed in purest hydrochloric acid. The solution was evaporated and the ammonium chloride sublimed. This was weighed and added to a solution of silver nitrate. The precipitate then was weighed.

To determine the weight of silver nitrate obtained from pure silver, Richards dissolved silver in nitric acid that had been repeatedly distilled and diluted with highly purified water. The solution was evaporated in a gentle current of air. The silver nitrate was finally dried and weighed.

Richards was awarded the Nobel medal in 1914, primarily for determining the numbers 107.881, 35.4574, and 14.0085 for silver, chlorine, and nitrogen. Once these weights were known, those of most of the other metallic elements could be determined, as the pure chlorides of the metals could be prepared and AgCl precipitated. For example, with a divalent metal, the reaction

$$MCl_2 + 2Ag^+ \rightarrow 2AgCl(s) + M^{++}$$

could be carried out and the AgCl weighed. The weight of AgCl obtained then allowed calculation of the atomic weight of M. The accuracy of all atomic weights determined in this way depends ultimately on the value used for silver.

In 1929 Giauque and Johnson discovered that oxygen atoms are not all identical, but vary in mass according to the number of neutrons in the nucleus. Thus, oxygen atoms of mass 17 and 18 (isotopes) as well as 16 exist. Subsequently, the standard of oxygen equal to exactly 16 became unsatisfactory, and dissatisfaction grew more and more acute as natural variations in the isotopic composition of oxygen were uncovered that made its atomic weight uncertain by 1 part in 10,000. Since the atomic weight of other elements cannot be known more precisely than the standard to which they are compared, something had to be done. Part of the problem was that a standard had to be chosen that would be satisfactory to both chemists and physicists; also, any change should not be great enough to disrupt seriously the values already in use.

In 1959 and 1960, the International Union of Pure and Applied Chemistry (IUPAC) and the International Union of Pure and Applied Physics (IUPAP) agreed that the standard should no longer be natural oxygen equal to 16. Instead, the isotope carbon-12 was assigned a mass of exactly 12.0000, and all weights are now referred to this standard.

Fortunately for chemists, this agreement changed the atomic weights formerly in use by only about 1 part in 10,000, and so did not cause extensive disruption.

A committee of IUPAC meets periodically to evaluate the latest work on atomic weights and publish an updated table of the most reliable values. Although we can never know true atomic weights, the best values are becoming more and more precise, and changes are becoming smaller. For some elements, especially sulfur and boron, natural variations in isotopic composition limit the precision of reported atomic weights. The most precise determinations of atomic weights now involve measurements of isotope ratios by mass spectrometry rather than by chemical methods. For example, in 1960 the U.S. National Bureau of Standards determined the atomic weight of silver by mass spectrometry to be 107.8731 ± 0.0018. Table 5-2 compares several values obtained by early workers with current recommended values.

SELECTED REFERENCES

T. W. Richards, *Chem. Rev.* **1**, 1 (1925). A clear description of the classic work of Richards and his coworkers.

A. F. Scott and W. R. Ware, *J. Amer. Chem. Soc.* **79**, 4253 (1957). Description of the chemical determination of the atomic weight of fluorine by hydrolysis of perfluorobutyryl chloride and precipitation of the released chloride with silver nitrate.

A. F. Scott and M. Bettman, *Chem. Rev.* **50**, 363 (1952). A comparison of the chemical and physical atomic weight values for the monoisotopic elements.

F. Szabadvary, *History of Analytical Chemistry*, Pergamon, London, 1966, pp 139-144. A description of the early work of Berzelius on atomic weights.

W. Timmermans, *J. Chem. Educ.* **15**, 353 (1938). A short biography of Jean Servais Stas of Belgium.

5-2   DETERMINATION OF CHLORIDE AS SILVER CHLORIDE

> Every substance must be assumed to be impure, every reaction must be assumed to be incomplete, every method of measurement must be assumed to contain some constant error until proof to the contrary can be obtained. As little as possible must be taken for granted.
>
> *T. W. Richards (from Nobel Award speech, 1914)*

Background

This experiment provides practice in handling a colloidal precipitate and in using a filter crucible. The use of such a crucible permits drying of the precipitate at a low temperature; the temperature required to burn off conventional filter paper causes appreciable decomposition of many precipitates, including silver chloride.

In this experiment, chloride in an unknown sample is determined by addition of a dilute solution of silver nitrate to a solution of the unknown to precipitate silver chloride.

$$Ag^+ + Cl^- \rightleftarrows AgCl(s)$$

Precipitation is carried out from a solution slightly acidified with nitric acid. The precipitate is filtered in a Gooch crucible, washed, dried at a low temperature, and weighed as silver chloride.

Silver chloride is light sensitive; the reaction is

$$2AgCl \rightarrow 2Ag(s) + Cl_2$$

If this reaction takes place before filtration, additional precipitate is formed and the result will be high:

$$Cl_2 + 3H_2O \rightarrow H^+ + Cl^- + HClO$$

$$Ag^+ + Cl^- \rightarrow AgCl(s)$$

After filtration, chlorine is lost and the result is low. Avoid exposing the precipitate to strong light, particularly direct sunlight, but do not be concerned about normal laboratory illumination, which causes only a slight darkening.

Since silver chloride is a colloidal precipitate, the particles must be made to clump together so that they can be retained by a filter. Agglomeration is brought about by the presence of an electrolyte, in this case by silver nitrate and nitric acid. The wash water also must contain an electrolyte so that the silver chloride precipitate does not peptize during washing. A volatile electrolyte, nitric acid, is used to prevent the addition of weight to the precipitate.

Silver chloride has a significant temperature coefficient of solubility: 0.9 mg/liter at 10°, 1.7 mg/liter at 25°, 5.2 mg/liter at 50°, and 21 mg/liter at 100°C. For this reason it is filtered at or below room temperature.

With proper precautions, the determination of chloride can be carried out with exceptional precision; the stoichiometry is exceedingly well behaved. Accordingly, the reaction of halogens with silver is used extensively for halogen determinations in organic elemental analysis, and in inorganic analysis it is used for oxyhalides as well as for halides. In fact, the properties of silver chloride as an analytical precipitate are so superior that it has a prominent historical position in the determination of atomic weights (Section 5-1).

Procedure (median time 4 hr)

Prepare three Gooch crucibles[1] and dry in an oven at 115° C to constant weight. Dry the unknown sample in a weighing bottle at 110°C for at least 1 hr. Accurately weigh 0.2-g samples of the unknown into 250-ml beakers. Dissolve in 125 ml of distilled water and add 2 ml of 6 M HNO_3. Calculate the volume of 0.1 M $AgNO_3$ solution needed to precipitate the chloride in the unknown sample, assuming the sample to contain as much chloride as would pure NaCl. Add slowly with stirring this volume plus an excess of about 5 ml. Heat the solution to 90 to 95°C to coagulate the precipitate. Allow the precipitate to settle, and test for completeness of precipitation by adding about 1 ml of 0.1 M $AgNO_3$ solution to the supernate. When precipitation is complete, cover the beaker with a watch glass, and allow the precipitate to digest for at least 1 hr.[2] Decant the supernatant liquid through the Gooch crucible with

---

[1] If a filter pad of glass fiber is used, place a pad in the crucible. With the suction on, wet it with water. Wash with a few milliliters of 0.1 M nitric acid to remove possible acid-soluble components, and then wash with water.

[2] No harm is done if the precipitate stands overnight or until the next laboratory period prior to filtration, but the beaker should be protected from light, dust, and HCl fumes.

suction, and wash the precipitate in the beaker twice by adding small portions of 0.01 M HNO_3, swirling, and decanting into the crucible. Transfer the precipitate to the crucible and continue washing until a test of the filtrate for silver ion is negative. Test by collecting a 5- to 10-ml portion in a test tube suspended with a thread in the filter flask and adding a few drops of NaCl solution. Finally, wash the precipitate with two or three portions of water to remove HNO_3. Dry to constant weight at 115 to 125°C.

Report the percentage of chloride in the sample.

PROBLEMS

5–1.   A 0.2312-g sample of an unknown was analyzed for chloride by precipitation and weighing of silver chloride. If 0.2030 g of AgCl was obtained, what was the percentage of chloride in the sample?

5–2.   What volume of 0.1 M $AgNO_3$ would be required to just precipitate all the chloride in 0.35 g of NaCl?

5–3.   If a precipitate of AgCl weighing 0.5000 g is washed with 200 ml of 0.01 M HNO_3 at 25°C, what percentage of the precipitate is lost through solubility? With the assumption that the solubility of AgCl in 0.01 M HNO_3 is the same as in pure water, what percentage is lost if the temperature of the wash solution is 50°C? How much wash solution can be used at 25°C if the loss is to be held to 0.1%?

5–3   DETERMINATION OF NICKEL AS NICKEL DIMETHYLGLYOXIME[1]

Background

In the gravimetric determination of a substance, the precipitating reagent should be as specific as possible for that substance. Although no known reagent is completely specific for a single element, for nickel the compound dimethylglyoxime (DMG) comes close to this ideal. In ammoniacal solution, nickel is precipitated quantitatively by DMG as a bright strawberry-red complex, $Ni(DMG)_2$. Palladium is precipitated as a yellow complex, $Pd(DMG)_2$, from acid solution, but in the presence of ammonia it is held in solution as a stable ammine complex. Of the other metals only copper in high concentrations and mixtures of cobalt(II) and

---

[1] We prefer this simpler name to the more accurate bis(dimethylglyoximato)nickel(II).

iron(III) cause interference by precipitating along with the nickel. Tartrate is added to complex any iron or aluminum present that otherwise would precipitate as the hydroxide when the solution is made alkaline.

The specificity of dimethylglyoxime arises from the unusual nature of the nickel chelate. Two dimethylglyoxime ligands are coordinated to each nickel ion, with the nickel and all four coordinating nitrogen atoms in the same plane.

Dimethylglyoxime (DMG)   Ni(DMG)$_2$

The complex is termed *square planar* because the coordinating nitrogen atoms all lie in a plane at the corners of a square, with the metal in the center. The structure of these crystals is unusual in that the nickel atoms are stacked one above another and form weak bonds. This metal–metal bonding apparently provides sufficient stability in the crystal to give low solubility relative to other metal ions.

Nickel dimethylglyoxime is bulky when first precipitated and also tends to creep up the walls of the vessel containing it. For these reasons, the amount precipitated should be limited to no more than about 50 mg. The precipitate is stable and readily dried at 110 to 130°C; it begins to decompose at about 180°C.

A number of other organic reagents have been developed for use as analytical precipitants, among them 8-hydroxyquinoline, cupferron, and 1-nitroso-2-naphthol. With all three a hydrogen ion (or other cation) is

8-Hydroxyquinoline   Ammonium   1-Nitroso-2-Naphthol
(oxine)   Nitrosophenyl-
                     hydroxylamine
                     (cupferron)

removed from the ligand upon chelation with the metal to yield a neutral species insoluble in water. Thus 8-hydroxyquinoline (HOx) gives a precipitate with aluminum having the composition $AlOx_3$. With magnesium the precipitate is $MgOx_2 \cdot 2H_2O$, the last two coordination positions around magnesium being occupied by water molecules because in this case two ligands are sufficient to provide a complex of zero charge.

Cupferron is used primarily to separate iron, vanadium, and titanium from copper, lead, aluminum, nickel, and zinc. The precipitation is carried out in about 3 M hydrochloric or sulfuric acid. The precipitates are rarely weighed directly, but are ignited and weighed as the oxides. A convenient and rapid method of separation prior to analysis is extraction of the complexes from aqueous acid into another solvent such as chloroform.

1-Nitroso-2-naphthol precipitates cobalt as a 3:1 complex, $Co(C_{10}H_6ONO)_3$. The method is used principally to separate cobalt from nickel. The precipitate, which is always contaminated with decomposition products, either is ignited to the oxide, Co_3O_4, and weighed, or is redissolved and the cobalt determined by oxidation-reduction titration or by electrodeposition.

Procedure (median time 4.5 hr)

Prepare three Gooch crucibles for the filtration step by placing a borosilicate-glass filter pad in each. Hold each pad to the light to check for pinhole leaks before putting it in the crucible. Place each crucible in a suction filtration assembly, wet the pad with water, and then draw a few milliliters of 0.1 M HNO_3 through it to remove acid- or water-soluble substances. Finally, draw a few milliliters of water through the pad, and dry the crucible plus pad in an oven at 110 to 150°C until constant weight (± 0.2 mg) is reached.

Accurately weigh samples of 1 to 2 g into 400-ml beakers,[2] add 10 ml of 6 M HNO_3 to each, cover with a watch glass, and place on a hot plate to dissolve. When dissolution is complete, boil the solution for a short time to remove most of the nitrogen oxides. Dilute each sample to about 100 ml with water. Add about 0.5 g of tartaric acid and 5 drops of methyl red indicator to each sample, and then add 6 M NH_3 until the indicator just changes color.[3] Heat the solution to about 70°C, and slowly add 20 ml of 2% solution of sodium dimethylglyoxime (NaDMG). Stir, and immediately add 3 M $NH_4C_2H_3O_2$ dropwise until a permanent red precipitate of $Ni(DMG)_2$ forms. Slowly add another 20 ml in excess. Allow the precipitate to coagulate on the hot plate for at least a half hour.

Add 1 to 2 ml of NaDMG solution to the supernatant liquid to test for completeness of precipitation. When complete, add about 10 drops of

---

[2] This size of sample is appropriate for materials that contain approximately 1 to 5% nickel.
[3] If the indicator fades before a permanent color change is reached, add more indicator and continue.

a 10% solution of detergent and allow to cool to room temperature before filtering. If precipitation is not complete, add about 10 ml more of NaDMG solution, stir, and allow to digest at 70°C for another 30 min. Decant the solution through the Gooch crucible with minimum disturbance of the precipitate. Again test the cold filtrate for completeness of precipitation. Transfer the precipitate, and wash it with small portions of a cold solution containing 1 ml of $6\,M$ NH_3 and a few drops of liquid detergent in 500 ml of water. Dry the crucible plus precipitate to constant weight (again ± 0.2 mg) in an oven at about 110°C.

Report the percentage of nickel in the sample. The formula weight of the nickel dimethylglyoxime precipitate, $Ni(DMG)_2$, is 288.93.

PROBLEMS

5–4.   A 1.7462-g sample of an alloy was analyzed by the procedure of this experiment. The weight of the empty crucible was 16.7176 g, and that of the crucible plus nickel dimethylglyoxime precipitate was 17.7348 g. What was the percentage of nickel in the sample?

5–5.   A 0.4767-g sample of an alloy containing only cobalt and nickel gave a nickel dimethylglyoxime precipitate weighing 0.8102 g when analyzed by the above procedure. What was the percentage of cobalt in the sample?

SELECTED REFERENCE

H. Diehl, *The Applications of the Dioximes to Analytical Chemistry,* G. F. Smith Chemical Co., Columbus, Ohio, 1940.

5–4   DETERMINATION OF SULFATE AS BARIUM SULFATE

Background

In this experiment the sulfate in a sample is determined by precipitation as barium sulfate. Precipitation is accomplished by slow addition of a dilute solution of barium chloride with stirring to a hot hydrochloric acid solution of the sulfate salt. The barium sulfate is then digested, filtered, dried, and weighed. For high-quality results all operations must be carried out so that no more than a few tenths of a percent of the precipitate is lost.

Barium sulfate is particularly susceptible to coprecipitation of foreign ions, and so care must be taken to keep the supersaturation ratio as low as possible during precipitation.

Attention to several other details in addition to the supersaturation ratio helps provide quantitative results. These include the following:

1. Hydrochloric acid added to the sulfate solution prevents the precipitation of barium salts of weak acids, such as carbonate or phosphate, and replaces adsorbed impurities. The acid volatilizes during the ignition step.
2. The presence of a slight excess of barium chloride lowers the amount of sulfate remaining in solution by the common-ion effect.
3. Digestion of the precipitate increases the average particle size and further reduces coprecipitation. In this operation a precipitate is allowed to stand in contact with the solution from which it was precipitated (mother liquor). For barium sulfate adequate digestion requires at least a half hour near the boiling point or several hours at room temperature.

Procedure (median time 4.2 hr)

Prepare a set of Gooch crucibles by placing a borosilicate-glass filter pad in each. Examine the pads individually for holes by holding them to the light. Place each crucible in a suction filtration assembly, apply suction, and wet the pad, first with water and then with a few milliliters of 0.1 M HNO_3 to remove soluble material. Set each inside a large conventional crucible, position over a Bunsen burner on an iron ring and clay triangle, and ignite for 10 to 15 min with a burner flame adjusted so that the bottom of the outer crucible is faintly red.[1] Allow to cool to room temperature (½ to 1 hr) and weigh the Gooch crucible. Repeat the nitric acid rinsing and drying until each crucible comes to constant weight, that is, within 0.2 mg of the previous weight. Each heating and cooling cycle takes about 1 hr.

Weigh 0.5-g portions of a dry sulfate sample into clean, but not necessarily dry, 400-ml beakers and dissolve in 250 ml of distilled water. Add 3 ml of 6 M HCl. Heat the solution to nearly boiling. Add slowly an excess of 0.05 M $BaCl_2$ solution with efficient stirring (about 70 ml is required). Use a separate stirring rod without policeman for each sample.

---

[1] If the crucible is ignited too strongly, some of the glass fiber mat melts, particularly near the edge. The maximum temperature should not exceed 500°C, where a porcelain crucible will just begin to glow red.

Allow the covered precipitate to digest near the boiling point for 30 min on a hot plate or over a low flame. Alternatively, allow to stand for at least several hours at room temperature. Test for completeness of precipitation by adding a few more drops of $BaCl_2$ solution to the liquid above the precipitate. If cloudiness appears, indicating that some sulfate is still in solution, add another 5 to 10 ml of $BaCl_2$ solution and digest again.

When ready to filter, place a crucible in the filtration assembly and decant the clear, hot, supernatant liquid through the filter. With a policeman loosen the precipitate from the beaker walls and the stirring rod; then transfer the precipitate quantitatively to the crucible with the aid of a wash bottle having an upward-pointing delivery tube. Wash the precipitate and crucible walls with several small portions of water, allowing each portion to drain completely before adding the next. After washing, test a portion of the filtrate for complete removal of soluble salts by checking for chloride. Do this by collecting a 5- to 10-ml portion in a test tube suspended by a thread in the filter flask and adding 1 ml of 0.1 M HNO_3 and a few drops of $AgNO_3$ solution. Ignite the crucible to constant weight as before. It is unnecessary to wash the precipitate between weighings.

Calculate and report the percentage of sulfur in the sample.

Calculations

To obtain the weight of sulfur, multiply the weight of $BaSO_4$ precipitate by the factor (at. wt S/ mol wt $BaSO_4$). In high-quality work the relative range for the results should not exceed 0.4%.

PROBLEMS

5–6.   A 0.4892-g sample of impure ammonium sulfate gave 0.8462 g of barium sulfate. What was the percentage of sulfur in the sample? What was the purity of the ammonium sulfate?

5–7.   A 0.3054-g sample of the mineral chalcopyrite ($CuFeS_2$) gave 0.6525 g of barium sulfate. What was the percentage of $CuFeS_2$ in the sample?

5–8.   A 0.5521-g sample of a sulfate salt contained 22.11% sulfur according to analysis by barium sulfate precipitation. Later it was found that the precipitate had been dried at only 110°C and contained 0.8% water at the time of weighing. What was the true percentage of sulfur in the sample?

5–9.   The solubility of barium sulfate in water is 4.4 mg/liter at 100°C. What volume of boiling wash water could be used to wash a 0.5432-g barium sulfate precipitate before the solubility loss would amount to 0.10%? Assume the wash water to be saturated with barium sulfate upon leaving the filter.

5–10.   What volume of 0.5 M $BaCl_2$ should be added to a sample containing 0.30 g of $(NH_4)_2SO_4$ to precipitate the sulfate as $BaSO_4$ and provide a 10% excess of reagent? What would be the weight of $BaSO_4$ precipitate obtained?

SELECTED REFERENCES

I. M. Kolthoff, E. B. Sandell, E. J. Meehan, and S. Bruckenstein, *Quantitative Chemical Analysis,* 4th ed., Macmillan, New York, 1969, pp 198, 228, and 602.

H. A. Laitinen and W. E. Harris, *Chemical Analysis,* 2nd ed., McGraw-Hill, New York, 1974, Chapters 7, 8, and 9.

D. G. Peters, J. Hayes, and G. Hieftje, *Chemical Separations and Measurements,* W. B. Saunders Company, Philadelphia, 1974, Chapter 8.

Chapter 6

OPTICAL METHODS OF ANALYSIS

One machine can do the work of one hundred ordinary men; no
machine can do the work of one extraordinary man.

Anon.

6–1 GENERAL BACKGROUND

The experiments of the preceding three chapters are designed to
develop the basic quantitative techniques of weighing, measuring volumes,
and making quantitative transfers of liquids and solids; all are operations
used extensively in most types of analysis. In the techniques of
measurement and separation introduced in this and subsequent chapters,
the experience gained thus far in the basic operations is used again and
again. Although the remaining experiments make use of a variety of
instruments, the emphasis is placed on the measurement or separation step
and not on the fundamentals of instrument design. The theory and design
of instrumentation is an important field in its own right and is not
pursued here.

Instruments are used for routine analysis whenever possible because
they are well suited to repetitive operations, especially when the
composition of the samples does not vary widely. They often may be
partly or completely automated, so that the analyst is concerned chiefly
with providing the necessary reagents and solutions, running standards as
checks or for calibration, and recording and interpreting the results. A
knowledge, then, of how to obtain and handle data from the basic types
of instruments is important to anyone making analytical measurements or
using output from instruments.

Analysis by Light Absorption

The determination of a trace of ammonia is the first of several
experiments designed to illustrate the principles and techniques involved

in the use of radiation-absorption measurement as an analytical tool. The measurement of color intensity as a means of analysis is characterized by speed, simplicity, and high sensitivity. These basic advantages have made photometry the most widely applied analytical method even though the precision is usually low. The apparatus employed range from simple glass tubes to complex instruments costing tens of thousands of dollars.

A minimum of three components is required for measuring color intensity: a light source, a sample holder or cell, and a detector for the unabsorbed light. The devices can be classified in various ways. One type depends on a comparison or matching of colors by one of several techniques, and another on direct measurement of the light-absorbing properties of a system. Instruments at several levels of sophistication are available, and for a given analysis a number of different types of instruments can be employed.

Since only certain discrete energy levels exist for the electrons in atoms or molecules, energy is absorbed or emitted only at wavelengths corresponding to electron transfers between these levels. The relation between energy absorbed or emitted E, wavelength λ, and frequency ν is given by

$$E = h\nu = hc/\lambda \qquad (6-1)$$

where h (Planck's constant) is equal to 6.63×10^{-27} erg-sec and c, the velocity of light, is equal to 3.0×10^{10} cm/sec. Therefore E is in units of ergs.

The fundamental expression for the amount of radiant energy absorbed by a solution is

$$\log \frac{I_O}{I_T} = abc = A = \log \frac{100}{\%T} \qquad (6-2)$$

where I_O is the intensity of incident radiation, I_T the intensity of emergent radiation from the sample, a a constant, b the length of sample in the radiation path, and c the concentration of the absorbing species in the sample. This equation is known as the *Beer-Lambert law*. If b is in centimeters and c in moles per liter, the constant a is called the *molar absorption coefficient* or molar absorptivity. Sometimes the ratio I_T/I_O, called *transmittance T*, is used. The term $\log(I_O/I_T)$ is given a special symbol, A, and is called *absorbance*. It can be seen from Equation (6-2) that absorbance is directly proportional to concentration, provided the length of sample in the light path is held constant. This equation is applicable to analytical measurements in all frequency ranges and for all samples that absorb radiant energy. Its principal quantitative use, however, is in the ultraviolet, visible, and infrared spectral regions, where energy absorption corresponds primarily to electron transfer between energy levels or to molecular vibrations or rotations.

In general, spectral measurements may be classified as either emission, in which the sample upon excitation emits radiation whose intensity is proportional to the concentration of the species present, or absorption, in which light from an independent source is passed through a sample and the amount that emerges is measured. The samples may be solids, liquids, or gases; a variety of cells for containing liquids and gases during spectral measurements are available.

Preliminary experimental operations for the measurement of light absorption in the visible and ultraviolet regions resemble those in volumetric analysis. Samples and standards are prepared, and normally aliquots are taken prior to the final measurement step. Since the measurement of radiant energy is related to concentration, the principles underlying light absorption are the main new aspects to be considered.

6–2   BASIC COLORIMETRY USING A STANDARD SERIES: AMMONIA BY THE NESSLER METHOD

> The best education is to be found in gaining the utmost information from the simplest apparatus.
>
> *A. N. Whitehead*

Background

The determination of ammonia outlined here employs a comparison method of measurement, the standard-series technique. This is one of the oldest and most reliable of all methods of color measurement. For ammonia it is fast, simple, and extremely sensitive. The precision, however, is poor; 1 part in 20 is about as much as can be expected. The light source is ordinary laboratory illumination, and the detector is the eye. The technique is therefore inexpensive and straightforward.

Precision in Measurement

In laboratory work, the precision needed for each operation always must be recognized. Since in this experiment the precision expected is only about 1 part in 20, each measurement need be made to only a few parts per hundred. To apply corrections to pipet volumes, for instance, is unnecessary, and the weighing of ammonium chloride for the standard solution of ammonia need be done only to the nearest milligram.

Reagent Impurities in Trace Analysis

A serious difficulty in trace analysis is the presence of interfering impurities at the trace level. For example, in the determination of ammonia at the level of parts per million, even small amounts of ammonia or ammonium salts in the distilled water or sodium hydroxide can lead to high results. This problem is especially serious when trace amounts of common substances such as iron or carbon dioxide are being determined. One or both of two procedures may be followed to reduce the determinate error. The first and most satisfactory is to use only reagents in which the concentration of substance of interest is too low to interfere significantly. If this procedure is too costly or time consuming, then the amount of sought-for substances in the reagents should be taken into account by a blank determination, in which all the reagents are present but the sample is not. This procedure is acceptable whenever the quantity of impurities is not a significant fraction of the total amount of the sought-for material.

The Nessler Method for Ammonia

The method for the determination of ammonia discussed here is a modification of the original procedure developed by Nessler in 1856. The color-forming reagent, Nessler's reagent, is prepared by first dissolving potassium iodide and mercury(II) chloride in a solution containing enough mercury(II) to just begin precipitating mercury(II) iodide:

$$HgCl_2 + 4I^- \rightleftarrows HgI_4^= + 2Cl^- \tag{6-3}$$

The solution is then made strongly alkaline with potassium or sodium hydroxide. The determination of ammonia is based on the reaction between this alkaline mercury(II) iodide solution and an alkaline solution containing ammonia or ammonium salts.

$$2HgI_4^= + 2NH_3 \rightleftarrows NH_2Hg_2I_3 + 5I^- + NH_4^+ \tag{6-4}$$

The ammonium ion produced reacts with more base to produce more ammonia and form additional $NH_2Hg_2I_3$. The insoluble $NH_2Hg_2I_3$ appears as an intense yellow colloid that is the basis for the measurement. Since the system is not a true solution, the relation between absorbance and concentration is not linear; under these conditions the standard-series technique is most efficient for the measurements. The color comparison is made in long, thin, flat-bottomed tubes called Nessler tubes. The final comparison with a set of standards closely spaced in concentration can be made readily in a special rack for holding the tubes.

Because the sensitivity of the Nessler technique is so high, ordinary water or distilled water is far too impure to use in the preparation of standards and reagents. Obtaining ammonia-free water formerly was the most onerous part of the experimental work, but now is accomplished simply by passing distilled water through a bed of strong-acid cation-exchange resin in the hydrogen form.

Determination of traces of ammonia is important in the field of water pollution. The ammonia content of water supplies is a reliable index of suitability for general use. The technique can be used for measurement of traces of ammonium salts or ammonia in gases. It also can be used to measure traces of protein nitrogen by analysis of the distillate from an alkaline permanganate distillation of the protein-containing material. Traces of nitrates or nitrites can be measured by reduction with suitable alloys, followed by distillation and measurement of the ammonia formed.

Procedure (median time 3.0 hr)

Preparation of Standard NH_4Cl Solution

Obtain ammonia-free water by passing distilled water through a cation-exchange column.[1] Test a sample of the water with Nessler's reagent before use by adding 10 ml of 2.5 M NaOH to a Nessler tube, diluting to the mark with a portion of the water, mixing, then adding about 4 ml of Nessler's reagent as described under Procedure for Analysis.[2]

Dissolve 0.250 g of NH_4Cl in water and dilute to 50 ml in a volumetric flask. Pipet 10 ml of this solution into a 50-ml volumetric flask and dilute to volume. Similarly, pipet 10 ml of the diluted solution into a 50-ml volumetric flask and dilute to volume with ammonia-free water. In the same way, pipet 10 ml of the doubly diluted solution into a 50-ml volumetric flask, and dilute to volume with ammonia-free water. The resulting standard solution should contain about 0.04 mg of NH_4Cl/ml.

Procedure for Analysis

Obtain an unknown sample and quantitatively transfer it to a 250-ml volumetric flask. Dilute to volume. Use ammonia-free water throughout.

---

[1] A drop of 3 M HCl (no more) added to each liter of distilled water before passage through the column ensures that any NH_3 present is converted to NH_4^+ and held on the resin.

[2] A yellow color may be due to contaminants in either the water or the equipment. Glassware must be rinsed thoroughly after cleaning because most synthetic detergents give a strong yellow color or form a precipitate with Nessler's reagent.

Also obtain five Nessler tubes, a rack to hold them for viewing, and a 2-ml pipet.

First Comparison. To a Nessler tube half filled with ammonia-free water add 1 ml of the sample, then 10 ml of 2.5 M NaOH solution (NH_3 free),[3] and dilute to volume with water.

Similarly, to four Nessler tubes add 0.3, 1.0, and 3.0 ml of standard. Add 10 ml of 2.5 M NaOH to each tube, dilute to volume, stopper, and mix by inverting each tube about five times. Add 2 medicine dropperfuls (~4 ml) of Nessler's reagent to each tube. The high density of the reagent makes mixing unnecessary. Allow the tubes to stand for 10 min to complete color development, and then visually compare the intensity of the color in the sample tube with that of the four standards. For easy comparison, place the standard tubes in alternate spaces in a Nessler rack; the sample tube then may be placed between the standards. The rack is used with the reflecting glass sloping away from the observer.

Second Comparison. Prepare standards as follows. To four Nessler tubes add 1.0, 1.3, 1.7, and 2.0 ml of standard. To each tube add 10 ml of 2.5 M NaOH, dilute to volume, mix, and then add Nessler's reagent as before.

Prepare sample as follows. To the fifth tube add 10 ml of 2.5 M NaOH and a predetermined volume of sample, selected in the following way. If the first comparison showed the ammonia concentration of the sample to be equivalent to:

a. less than 0.3 ml of standard, use 4 to 7 ml of sample in the second comparison;

b. between 0.3 and 1.0 ml of standard, use 2 to 3 ml of sample in the second comparison;

c. between 1.0 and 3.0 ml of standard, but nearer 1.0 than 3.0 ml, use 1 ml of sample in the second comparison;

d. between 1.0 and 3.0 ml of standard, but nearer 3.0 than 1.0 ml, dilute 20 ml of sample to 50 ml in a volumetric flask and use 1 ml for the second comparison;

e. more than 3.0 ml of standard, dilute 10 ml of sample to 50 ml in a volumetric flask. In this case repeat the first comparison, and on the basis of the results proceed to the second comparison. Remember to account for this dilution in the calculations.

Dilute the sample to volume in the Nessler tube, mix, and add Nessler's reagent as before. Visually compare the intensity of color with that of the four standards.

Third Comparison. Prepare a third set of standards. The lowest standard

---

[3] The NaOH solution should be prepared from solid NaOH and NH_3-free water. Stock solutions of NaOH provided in the laboratory frequently are not sufficiently free of ammonia for this experiment.

concentration should be that concentration whose color intensity is just below that of the sample in the second comparison. The highest standard concentration should be that concentration whose color intensity is just above that of the sample in the second comparison. Use the volume of sample that was used in the second comparison. Add NaOH and Nessler's reagent to the sample and standards as before. Small differences in color intensity can be seen more readily with natural than with artificial light.

Record the volume of standard whose color intensity most closely matches the color intensity of the sample solution.

Report the results as micrograms of ammonia per milliliter of solution in the 250-ml flask containing the sample. Neither averages nor medians are applicable in this experiment.

Calculations

Calculate the ammonia concentration in the sample from the experimental procedure as follows. When two Nessler tubes give matching color intensities, as in the final color comparison at balance,

$$\mu\text{g in sample} = \mu\text{g in standard} \qquad (6-5)$$

The standard contains an initial weight in grams of NH_4Cl in 50 ml that has undergone four dilutions (original plus three). The original 50 ml contains $(\text{mol wt } NH_3 \times \text{g } NH_4Cl)/(\text{mol wt } NH_4Cl)$, or $[(\text{mol wt } NH_3 \times \text{g } NH_4Cl \times 10^6)/(\text{mol wt } NH_4Cl \times 50)]\,\mu\text{g/ml}$. As a result of the dilutions the micrograms of NH_3 in the final comparison tube is given by

$$\mu\,NH_3 = \frac{NH_3}{NH_4Cl} \times \text{g } NH_4Cl \times \frac{10^6}{50} \times \frac{10}{50} \times \frac{10}{50} \times \frac{10}{50} \times V_{std}$$

where V_{std} represents the milliliters of standard present in the final color comparison. This also gives the micrograms of ammonia present in the final comparison tube that contains V_{smp} ml of the sample solution. Since V_{smp} ml of the sample was used, then for 1 ml of sample in the 250-ml volumetric flask the $\mu\text{g } NH_3$ is

$$\mu\text{g } NH_3/\text{ml} = \frac{1}{V_{smp}} \times \frac{NH_3}{NH_4Cl} \times \text{g } NH_4Cl \times$$

$$\frac{10^6}{50} \times \frac{10}{50} \times \frac{10}{50} \times \frac{10}{50} \times V_{std} \qquad (6-6)$$

$$= \frac{V_{std}}{V_{smp}} \times \frac{NH_3}{NH_4Cl} \times \text{g } NH_4Cl \times 160$$

PROBLEMS

6–1.   A 5.0-ml sample treated by the procedure of this experiment gave a color intensity that matched a tube containing 1.2 ml of standard ammonium chloride solution (0.263 g NH_4Cl/50 ml). What was the NH_3 concentration in the sample in micrograms per milliliter?

6–2.   A 100-ml sample of Lake Ontario water in a Nessler tube gave the same color intensity as 1.3 ml of a standard solution prepared from 0.261 g of ammonium chloride and diluted by the procedure of this experiment. How many micrograms of ammonia were in the sample? How many micrograms were present per milliliter?

6–3.   A sample of ammonium chloride weighing 0.4194 g was dissolved and diluted to 500 ml. A 25.00-ml aliquot was diluted to 1 liter and a 25.00-ml aliquot of that solution diluted to 250 ml. What was the NH_3 concentration of the final solution in micrograms per milliliter?

6–4.   A bottle of reagent-grade potassium chloride was tested for ammonia by the Nessler procedure. A solution containing 10.2 g gave the same color in a Nessler tube as 1.4 ml of a standard solution prepared from 0.286 g of ammonium chloride and diluted by the procedure of this experiment. What was the ammonium ion concentration in the potassium chloride in micrograms per gram?

SELECTED REFERENCES

D. F. Boltz, Ed., *Colorimetric Determination of Non-metals,* Wiley, New York, 1958.
J. Nessler, *Chem. Ztg.* **27** (N.F.I.), 529 (1856).

6–3 SPECTROPHOTOMETRIC ANALYSIS IN THE VISIBLE REGION: DETERMINATION OF TRACE IRON AS BIPYRIDINE COMPLEX

Background

In Section 6–1 the relation between absorption of radiant energy, thickness of an absorbing layer, and concentration of a colored species was introduced. The color intensity was estimated by eye. Since the human eye responds to energy in the spectral region between about 400 and 750 nanometers (millimicrons), any substance that absorbs light within this range interferes with the Nessler method for ammonia. Compounds that absorb radiant energy outside this range do not interfere because the eye does not respond to shorter or longer wavelengths.

High sensitivity is obtained in the Nessler method by use of a long column of liquid. Another way to increase sensitivity is to use a narrow band of radiation as a light source. For instance, if a material absorbs only yellow light, a light source emitting only yellow light is best. Incident radiation from other regions of the spectrum does not enhance, but reduces, the sensitivity. The wavelength region of the incident light can be partially restricted in a simple way with filters; several instruments have been designed incorporating them. A still narrower wavelength band can be provided by a monochromator.

Restricting the wavelength band incident on a sample has the additional advantage of increasing selectivity; species that absorb outside the incident band of radiation cannot interfere. Greater precision (up to about 1 part per 100) can be gained by substituting a photosensitive detector for the eye and a more reproducible light source for daylight. Furthermore, a reduction in the number of spectral interferences in an analytical method results in less need for separations.

Figure 6-1 shows the basic components of an instrument for measuring the radiant energy absorbed by a sample. Radiation from a stable source, usually a tungsten lamp in the visible region, passes through a slit to a monochromator, or wavelength selector, which allows only a narrow range of wavelengths to emerge through an exit slit. Generally this selector utilizes a prism or a grating to disperse the incident light. Rotation of the prism or grating causes the desired wavelength band to fall on the exit slit. The radiation then passes through the sample, and the

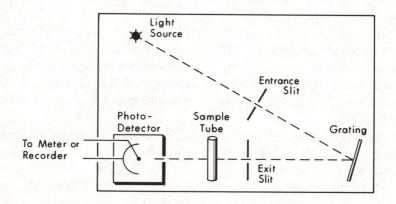

FIGURE 6–1. Schematic diagram of a grating spectrophotometer.

portion not absorbed is measured by a photodetector, which produces an electrical signal proportional to the radiation intensity.

The most important part of a spectrophotometer is the mono-chromator. It provides increased selectivity (freedom from interference) as well as increased sensitivity. A grating is used in the Bausch and Lomb Spectronic 20 instrument referred to in this experiment.

Bipyridine for the Analysis of Trace Iron

Bipyridine (bipy) forms an intensely red complex with iron(II) that may be used to determine iron concentrations in the range of parts per million. The reaction is

$$3\text{Bipy} + \text{Fe}^{++} \rightleftarrows \text{Fe(Bipy)}_3{}^{++} \tag{6-7}$$

The reagent and complex have the structures

The molar absorption coefficient (a in the Beer-Lambert law) of the complex is 8650 liters/mole-cm at 522 nm, the wavelength of maximum absorption. The complex forms rapidly, is stable over a pH range of 3 to 9, and may be used to measure iron(II) concentrations in the range of 0.5 to 8 ppm. Iron(III), if present, must be reduced to iron(II) to produce the colored species. A suitable reagent for this purpose is hydroxylamine hydrochloride, $NH_2OH \cdot HCl$.

Although the concentration of iron in the sample could be calculated from the Beer-Lambert law if the molar absorption coefficient and solution thickness were known, it is preferable to prepare one or more standards and compare absorbance readings of the sample and standard solutions. In this way the effects of instrument and solution variations are minimized.

Spectrophotometric methods are normally accurate to about 1%, although higher accuracy and precision can be attained with more sophisticated instruments and special techniques. In most cases 1% accuracy at the milligrams-per-liter level is sufficient. The standard used in this experiment, $FeSO_4(NH_4)_2SO_4 \cdot 6H_2O$, although not a primary standard, has a purity greater than 99%, which is adequate.

Sample Tubes (Cuvettes) in Spectrophotometry

Several different types of sample containers are used in spectrophotometry. Less expensive instruments are designed to use test tubes for liquid samples. To ensure that the solution path length is the same for sample and standard solutions, a matched set of tubes is needed; the use of a single tube for all measurements, although possible, is inconvenient. Tubes are matched by placing a solution of intermediate absorbance in each and comparing absorbance readings. One tube is picked arbitrarily as a reference, and others are selected that give the same reading within 1%.

Procedure (median time 2.7 hr)

*Selection of Matched Tubes for Bausch and Lomb
Spectronic 20 Spectrophotometer*

Obtain a supply of 13- by 100-mm test tubes that are clean, dry, and free of scratches. Put an index mark near the top of each. Half fill each tube with a solution containing 2 g of $CoCl_2 \cdot 6H_2O$ in 100 ml of 0.3 M HCl.[1] Set the wavelength to 510 nm (mμ) on the spectrophotometer. (See operating instructions for the Bausch and Lomb Spectronic 20 at the end of this experiment.) With the amplifier control, set the instrument to read zero. Place a tube in the sample compartment, and adjust the light control so that the meter reads 90% transmittance. Using this tube to check periodically the 90% reading, insert the other tubes and record their transmittance. Insert each tube in the same position relative to its index mark. Choose a set of seven tubes that have less than 1% variation in reading. (Three tubes are sufficient if Experiment 6-5, cobalt–nickel, is not performed.) Retain these tubes for subsequent photometric work, and return the remainder. To compensate for variations in instruments, use the same instrument for both tube matching and experimental work.

Preparation of Standard Iron Solution

Weigh enough $FeSO_4 \cdot (NH_4)_2 SO_4 \cdot 6H_2O$ to prepare 250 ml of a solution 0.002 M in iron. Transfer the salt to a 250-ml volumetric flask, dissolve in water, add 8 ml of 3 M H_2SO_4, dilute to volume with distilled water, and mix. Pipet 10 ml of this solution into a 100-ml volumetric flask, add 4 ml of 3 M H_2SO_4, and dilute to volume.[2]

---

[1] This solution is selected because it is stable, has a broad absorption band at about the center of the visible region, and transmits about 50% in a 1-cm cell.

[2] The iron concentration of this standard solution should be known to within about 0.5%.

Analysis Procedure

The procedure outlined here is for a liquid sample of, say, ground or stream water. Transfer the entire sample to a 100-ml volumetric flask, add 4 ml of $3 M H_2SO_4$, mix, dilute to volume, and mix again. Pipet duplicate 10-ml portions into two 50-ml volumetric flasks. Into another pair of flasks pipet duplicate 10-ml portions of the standard iron solution. For a blank add 0.4 ml (about 10 drops) of $3 M H_2SO_4$ to a fifth 50-ml volumetric flask.

Add, in order, to each of the five flasks, 1 ml of 10% hydroxylamine hydrochloride solution, 10 ml of 0.1% bipyridine solution, and 4 ml of 10% sodium acetate solution,[3] mixing after each reagent solution is added. Dilute to volume. Read several times the absorbance of each solution at 522 nm, using matched tubes. Set the absorbance reading to zero with the blank solution.

Report the total weight in milligrams of iron in the original sample.

Operating Instructions for Bausch and Lomb
Spectronic 20 Spectrophotometer

See Figure 6-2 for location of controls.

1. Turn on the instrument by rotating the amplifier control clockwise.
2. Set the wavelength control to the desired wavelength. With the amplifier control, set the meter needle to zero on the percent-transmittance scale. This setting corresponds to infinity on the absorbance (optical density on older instruments) scale.
3. Dry the outside of the matched sample tubes with a lintless towel or tissue. Insert a tube containing a blank into the sample compartment. Position the tube in the instrument with the aid of the index mark. Close the cover of the compartment.
4. Rotate the light control until the meter reads 100 on the percent-transmittance scale (zero on the absorbance scale). Figure 6-3 shows the function of this control.
5. Remove the blank tube and recheck the zero reading. Replace the blank tube with one containing a standard or sample. Read the absorbance directly and record. The cover of the sample compartment should be closed for all readings. Variations of ±1% in the readings are normal.

---

[3] The sodium acetate plus sulfuric acid gives an acetic acid–sodium acetate buffer in the pH region of about 4.5 to 5.

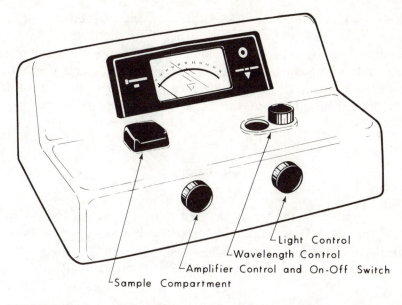

FIGURE 6-2. External view of Bausch and Lomb Spectronic 20 spectrophotometer.

6. If readings are to be taken at another wavelength, remove the sample tube and insert the blank tube. Turn the light control counter-clockwise before changing the wavelength setting; otherwise, increased sensitivity of the photodetector at the new wavelength may result in a signal of sufficient magnitude to damage the meter. Set the new wavelength.

7. Repeat Steps 4, 5, and 6 until readings at all desired wavelengths have been taken. Readjust the light control whenever the wavelength setting is changed. The zero percent-transmittance setting (dark current) also should be checked periodically and readjusted as needed with the amplifier control.

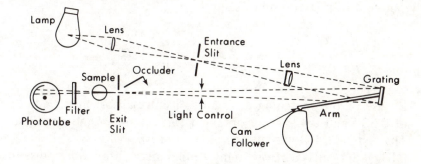

FIGURE 6-3. Diagram of the optical path of Bausch and Lomb Spectronic 20 spectrophotometer.

Readings taken at 10- to 20-nm intervals are sufficient to outline an absorption curve except at absorption peaks or shoulders, where additional points may be needed to characterize the curve more completely. Near the ends of the spectral range of the instrument, below about 350 and above 650 nm, a 100% transmittance reading may be impossible to obtain with the blank. In these regions use a lower reading.

Calculations

A comparison method may be used only if a system follows Beer's law. For the iron(II)–bipyridine system the law is followed over the concentration range of 0.5 to 8 ppm. In this range, therefore, the absorbance is proportional to the amount of iron present, and

$$\frac{A_{std}}{A_{smp}} = \frac{[Fe]_{std}}{[Fe]_{smp}} = \frac{(mg\ Fe/ml)_{std}}{(mg\ Fe/ml)_{smp}} \qquad (6-8)$$

where A_{std} is the average absorbance reading of the standard and A_{smp} is the average absorbance reading of the sample. The concentration of iron in the final standard solution, then, is given by

$$(mg\ standard)\ \frac{at.\ wt\ Fe}{mol\ wt\ FeSO_4(NH_4)_2SO_4 \cdot 6H_2O}\left(\frac{10}{250}\right) =$$

$$mg\ Fe\ in\ 100\ ml\ of\ standard\ solution \qquad (6-9)$$

Substitution of this value and the average absorbance readings for standard and sample into Equation (6-8) gives the total weight (in milligrams) of iron in 100 ml of sample.

Example 6–1.

By use of the foregoing procedure, average absorbance readings of 0.337 and 0.428 were obtained for a standard and a sample. The standard was prepared from 0.413 g of $FeSO_4$-$(NH_4)_2SO_4 \cdot 6H_2O$. How many milligrams of iron were present in the sample?

First calculate the concentration of iron in the standard solution:

$$\text{mg Fe} = \frac{\text{at. wt Fe}}{\text{mol wt } FeSO_4(NH_4)_2SO_4 \cdot 6H_2O} \ (413 \text{ mg})$$

$$= 58.9 \text{ mg Fe in 250 ml}$$

$$(58.9 \text{ mg}) \ \frac{10 \text{ ml}}{250 \text{ ml}} = 2.36 \text{ mg Fe in 100 ml}$$

From Equation (6-8)

$$\frac{A_{std}}{A_{smp}} = \frac{(\text{mg Fe}/100 \text{ ml})_{std}}{(\text{mg Fe}/100 \text{ ml})_{smp}}$$

$$(\text{mg Fe}/100 \text{ ml})_{smp} = \frac{0.428}{0.337} \ (2.36)$$

$$= 3.00 \text{ mg Fe}/100 \text{ ml}$$

PROBLEMS

6–5.   The iron in a sample of stream water was determined by the procedure of this experiment. The data recorded were: weight of iron(II) ammonium sulfate taken, 0.3151 g; absorbance reading of standard, 0.452; absorbance reading of sample, 0.513. What was the percentage of iron present in 100 ml of the stream water?

6–6.   A solution of 0.0040 M copper dithizone transmits 66% of the light at a particular wavelength in a cell of 1.5-cm thickness. What is the absorbance? What is the molar absorption coefficient? What is the concentration of a copper dithizone solution that transmits 30% under the same conditions?

6–7.   A 2.0 × 10^{-4} M solution of a new compound gives a transmittance of 48.3% in a cell having a path length of 2.0 cm. What is the concentration of a solution having a transmittance of 28.2% under the same conditions? What is the molar absorption coefficient of the compound?

6—8.   A colored substance S (mol wt 150) was found to have an absorption peak at 405 nm. A solution containing 3.03 mg/liter had an absorbance of 0.842 when examined in a 2.50-cm cell. What weight of S is contained in 100 ml of a solution that has an absorbance of 0.768 at 405 nm when measured in a 1.04-cm cell?

6—9.   The intensity of a monochromatic light beam is decreased to 18% of its original intensity on passing through 3.00 cm of a $2.14 \times 10^{-4} M$ solution of picric acid at a given wavelength. What is the molar absorption coefficient of picric acid at this wavelength?

6—10.   Explain briefly why spectrophotometric analyses are generally, but not always, carried out at the wavelength of maximum absorbance of the substance being determined.

6—11.†   If the solution used for the tube-matching procedure absorbs little light, tubes may appear to be matched when they are not. Why? Also, if the solution absorbs most of the incident light, matching will be poor. Why?

SELECTED REFERENCES

W. J. Blaedel and V. W. Meloche, *Elementary Quantitative Analysis,* 2nd ed., Harper & Row, New York, 1963, p 505.

M. L. Moss and M. G. Mellon, *Anal. Chem.* **14**, 862 (1942). First paper on determination of iron with bipyridine.

F. D. Snell, *Colorimetry,* 3rd ed., Van Nostrand Reinhold, New York, 4 vols., 1948-1954.

D. G. Peters, J. Hayes, and G. Hieftje, *Chemical Separations and Measurements,* W. B. Saunders Company, Philadelphia, 1974.

6—4 ATOMIC ABSORPTION SPECTROSCOPY: DETERMINATION OF TRACE COPPER IN NICKEL METAL

Background

Atomic absorption spectroscopy is a method particularly suited to the measurement of small amounts of elements, usually metals, in a sample. The element to be determined is dissociated from its environment so that it exists as free atoms in the ground state. Atoms in this state readily absorb electromagnetic radiation at wavelengths corresponding to excitation to higher energy levels. The extent of this absorption of energy from a radiation source is measured photometrically and compared with standard samples containing known amounts of the element. The method is unusually simple, sensitive, and selective.

An element is usually dissociated from its initial matrix by dissolution of the sample and aspiration of the resulting solution into a flame. The flame is located between a radiation source of optimum wavelength and a detector in a manner analogous to a spectrophotometer cell (Figure 6-4). The sample solution is first drawn into the burner–nebulizer by a stream of air or other oxidant such as nitrous oxide and then mixed with fuel (in Figure 6-4, acetylene). Most of the sample solution passes into the nebulizer chamber as relatively large droplets that settle out and pass down the drain tube. The remaining solution is carried by the air–fuel mixture as a mist to the burner head, where the solvent evaporates and the solute is dissociated into atoms by the heat of the flame. The number of atoms reaching this point in the operation constitutes only a small fraction of the total. For high sensitivity as many as possible of these atoms present in the flame should absorb radiation from the source. The ideal source for this purpose would be of high intensity at the wavelength needed for the element being determined but of low intensity at all other wavelengths. The nearest approach to this ideal is a lamp whose cathode contains the element being determined. Atoms of this element upon heating emit energy at the wavelengths most likely to be absorbed by that same element in the flame. In this way, selectivity as well as sensitivity is provided because other elements in the sample generally do not absorb close enough to the chosen wavelength to interfere in the measurement. Background interference is reduced by introduction of the monochromator (containing a quartz prism or, more commonly, a grating) between the sample and the detector.

Solutions more concentrated than 10^{-3} to 10^{-5} M should be diluted to bring the concentration into this range. Only a small volume of solution, on the order of a few milliliters, is required.

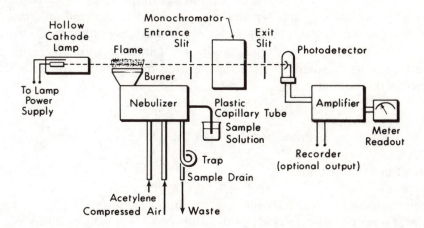

FIGURE 6–4. Schematic diagram of an atomic absorption spectrophotometer.

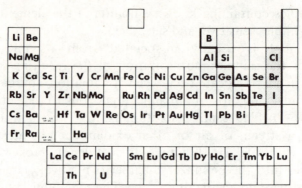

FIGURE 6–5. Elements determinable by atomic absorption. Shaded areas indicate elements determinable by indirect methods.

A large number of elements can be determined by atomic-absorption techniques (Figure 6-5). Many applications of atomic absorption to analysis of trace metals in a variety of organic, inorganic, and biological systems have been developed and are replacing slower, more tedious techniques. For example, atomic absorption has been found to be exceptionally satisfactory for the determination of magnesium in cast iron in the 0.002 to 0.1% concentration range, and of silver, zinc, copper, and lead in cadmium metal in the 0.004 to 0.4% concentration range.

Although atomic absorption has caught on quickly as a rapid, convenient analytical method, it must be used with care. The chief problems are instrument drift due to changes in lamp intensity or wavelength calibration with time and, because of the extreme sensitivity of the method, trace contamination from reagents, water, and surroundings during sample preparation and analysis. Frequent rechecks of instrument operation with standard solutions and care in the preparation and handling of both standard and sample solutions contribute to improved accuracy and precision of the technique. In this experiment, trace copper in a sample is determined by the method of *standard additions,* in which varying amounts of standard are added to a series of solutions containing a constant amount of sample. A plot of absorbance against amount of standard then can be used to determine the amount of copper in the sample.

Procedure (median time 1.8 hr)

Preparation of Solutions

Sample Solution. Weigh to the nearest milligram about 1 g of a sample into a 250-ml volumetric flask, add 20 ml of 6 M HNO$_3$, and dissolve by warming slightly on a hot plate. Cool, dilute to volume with water that has been passed through a cation or mixed-bed ion-exchange column to remove ionic impurities (particularly copper), and mix.

Standard Copper Solution. Prepare a concentrated copper solution by dissolving 0.150 g of pure copper metal in a 100-ml volumetric flask with 20 ml of 6 M HNO_3. Dilute to volume with deionized water and mix. Prepare a standard copper solution 100 times more dilute by serial dilutions of 10-ml portions to 100 ml.

Standard Solutions. With a buret, measure into six 50-ml volumetric flasks 0, 2, 5, 10, 15, and 20 ml of the standard copper solution. Pipet into each flask 10 ml of the sample solution. Dilute the contents of each flask to volume with deionized water and mix.

Take the six filled flasks, along with a 50-ml beaker and a wash bottle of deionized water, to an atomic absorption spectrophotometer, where instruction in the operation of the instrument will be provided. The optical-path diagram of a typical single-beam instrument is depicted in Figure 6-6.

Measurement of Absorption

Adjust the fuel and air settings for the burner to the proper values. After fuel and air flow is well established (about 5 sec), light the burner. *Caution:* If the mixture is not allowed to establish itself in the burner before ignition, a small explosion may occur inside the burner chamber. Never leave the flame unattended.

Aspirate deionized water from a 50-ml beaker (blank). Set the absorbance reading to zero.

Replace the blank with the most concentrated solution. Turn the control dial so that the instrument reads a specified value at the upper end of the absorbance scale (usually 100).[1]

Again aspirate the blank and concentrated solutions to ensure that the readings are reproducible.

---

[1] An unexpectedly low reading for a sample may be caused by a fouled burner slot, an obstruction in the sample capillary, or an incorrect wavelength setting. Notify the instructor if one of these problems is suspected.

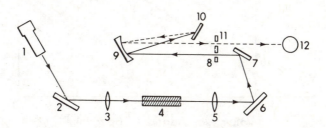

FIGURE 6–6. Diagram of optical path in the Perkin-Elmer Model 290B atomic absorption spectrophotometer: (1) hollow-cathode lamp; (2) mirror; (3) lens; (4) flame; (5) lens; (6) mirror; (7) mirror; (8) slit; (9) parabolic mirror; (10) grating; (11) slit; (12) photodetector. Components (8) through (11) make up the monochromator.

Aspirate the other solutions and record their readings.[2] Take each reading 5 to 10 sec after beginning aspiration. Repeat each aspiration and reading several times. Recheck the blank and concentrated-solution settings after every group of readings and reset if required.

Aspirate deionized water for a few minutes before shutdown to clean the capillary and chamber.

Report the percentage of copper in the sample. Report both the graphical and least-squares values, along with the standard deviation of the result calculated by the least-squares method. Indicate which value is to be graded.

Calculations

Using an 8½- by 11-in. sheet of accurately ruled graph paper, construct a plot of instrument readings (which are proportional to absorbance) against volume of copper solution. Draw the best straight line through the points. The plot will not go through zero on the vertical axis because of the copper from the sample present in each solution. The simplest way to measure the copper concentration of the sample is to extend the plot until it intersects the horizontal axis. If the horizontal axis is continued to the left of zero, the concentration of the sample can be obtained from the point of intersection. Figure 6-7 illustrates this method of plotting.

In calculating the percentage of copper in the original sample, recall that the sample was dissolved in a volume of 250 ml and that this solution

---

[2] The values may be read directly from a meter or a digital readout, or recorded on a strip-chart recorder.

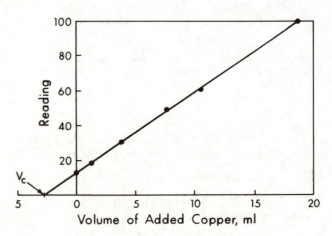

FIGURE 6–7. Example of plot of data obtained by method of standard addition. The intercept is proportional to the copper contributed by the sample.

was diluted fivefold to give the solution on which the measurements were actually made.

Calculation of Copper in a Sample by the Least-Squares Method

Measurements linearly related to concentration, and for which the slope or an axis intercept of the straight line needs to be determined, can be analyzed by least-squares calculation or by graphical methods. The mathematical method usually takes longer; it is generally more precise but not necessarily more accurate. The least-squares calculation can provide not only a value for the intercept but also the standard deviation of the intercept. This experiment affords an opportunity to compare these two methods of data evaluation. Carry out the graphical procedure first, so that the drawing of the line will not be prejudiced by prior knowledge of the least-squares results.

The equation for the volume of standard copper solution corresponding to the sample (the horizontal-axis intercept) is given by

$$V_c = \frac{(\Sigma A)(\Sigma V^2) - [\Sigma(AV)](\Sigma V)}{(n)[\Sigma(AV)] - (\Sigma A)(\Sigma V)} \qquad (6-10)$$

where V_c is the horizontal-axis intercept, ΣA the sum of the readings, ΣV^2 the sum of squares of the volumes of standard copper used, $\Sigma(AV)$ the sum of products of corresponding pairs of meter reading–copper volume sets, ΣV the sum of volumes of standard copper solution, and n the number of solutions measured. An example of a tabulation is given in Table 6-1.

TABLE 6—1. TABULATION OF DATA FOR LEAST-SQUARES CALCULATION

Volume of Standard Copper Solution, V	V^2	Average Reading, A	AV
0	0	13.0	0
2.03	4.12	21.5	43.64
4.99	24.90	35.0	174.65
9.99	99.80	56.5	564.44
15.01	225.30	77.0	1155.77
20.00	400.00	100.0	2000.00
ΣV 52.02	ΣV^2 754.12	ΣA 303.0	$\Sigma(AV)$ 3938.50

$$V_c = \frac{(303.0)(754.1) - (3938.5)(52.02)}{(6)(3938.5) - (303.0)(52.02)}$$

$$= \frac{228,498.7 - 204,880.8}{23,631 - 15,762} = \frac{23,617.9}{7869}$$

$$= 3.001 = 3.00 \text{ ml}$$

$$\text{Concentration of Cu standard} = \frac{0.100 \text{ g Cu}}{100 \text{ ml}} \times \frac{10 \text{ ml}}{100 \text{ ml}} \times \frac{10 \text{ ml}}{100 \text{ ml}}$$

$$= 1.00 \times 10^{-5} \text{ g/ml (for a standard of } 0.100 \text{ g)}$$

$$\text{g Cu from sample in solution} = (3.00 \text{ ml})(1.00 \times 10^{-5} \text{ g/ml})$$
$$= 3.00 \times 10^{-5} \text{ g}$$

Thus 10 ml of the sample solution contained 3.00×10^{-5} g of Cu. The original 250 ml, then, contained 7.50×10^{-4} g of Cu.

$$\% \text{ Cu in sample} = \frac{0.000750 \text{ g}}{1.0000 \text{ g}} (100)$$

$$= 0.075\% \text{ (for a 1.000-g sample)}$$

Calculation of Standard Deviation for Least-Squares Result

The error in the value of V_c determined above can be estimated as follows. First, calculate from the equation for the least-squares line an absorbance value A_{calc} for each value of V. Then the difference d between the calculated values and the measured values A are tabulated, and Σd^2 is determined. This quantity is then inserted in the equation for the variance of V_c, and the standard deviation is obtained. The steps are outlined below.

1.   Calculate the slope m of the least-squares line:

$$m = \frac{\Sigma A}{\Sigma V + nV_c} \tag{6-11}$$

2.   Calculate the values of A for each value of V:

$$A_{calc} = m(V + V_c) \qquad (6\text{--}12)$$

3.   Find values of d for each point:

$$d = A - A_{calc} \qquad (6\text{--}13)$$

4.   Find $\Sigma d^2/(n - 2)$, the variance of the readings, for the data.
5.   Calculate the variance of V_c:

$$\text{variance of } V_c = \frac{\Sigma d^2}{n - 2} \frac{\Sigma (V + V_c)^2}{(\Sigma A)^2 \left[\dfrac{n\Sigma (V + V_c)^2}{[\Sigma (V + V_c)]^2} - 1\right]} \qquad (6\text{--}14)$$

6.   Find the standard deviation of V_c:

$$\text{standard deviation} = \sqrt{\text{variance}} \qquad (6\text{--}15)$$

The standard deviation of V_c for the data in Table 6-1 is found as follows:

$$m = \frac{303.0}{(52.02) + (6)(3.00)} = \frac{303.0}{70.02} = 4.327$$

The data for the subsequent steps are summarized in Table 6-2.

$$\frac{\Sigma d^2}{n - 2} = \frac{1.462}{4} = 0.3581$$

TABLE 6–2. SUMMARY OF DATA FOR CALCULATION OF
STANDARD DEVIATION

V	$V + V_c$	A_{calc}	d	d^2	$(V + V_c)^2$
0	3.00	12.98	0.02	0.0004	9.00
2.03	5.03	21.76	0.26	0.0676	25.30
4.99	7.99	34.57	0.43	0.1849	63.84
9.99	12.99	56.21	0.29	0.0841	168.74
15.01	18.01	77.93	0.93	0.8649	324.36
20.00	23.00	99.52	0.48	0.2304	529.00
	70.02			1.4323	1120.24

$$\text{Variance} = (0.3581) \frac{1120.24}{91,809 \left[\dfrac{(6)(1120.24)}{4902.80} - 1 \right]}$$

$$= \frac{(0.3581)(1120.24)}{(91,809)(1.37094 - 1)} = 0.01178$$

$$\text{Standard deviation in } V_c = \sqrt{0.01178} = 0.109 = 0.11$$

The value for V_c, then, is 3.00 ± 0.11 ml. With the assumption that errors in the weights of the copper standard and the sample are negligible, the error in the percentage of copper reported is proportional to the error in the calculation of V_c. Therefore the answer and its uncertainty would be 0.149 ± 0.0055.

PROBLEMS

6–12.   A 0.9125-g sample of powdered nickel was analyzed for copper by the method of this experiment. The weight of copper used for the standard was 0.2214 g. The calibrated volume of the pipet used was 9.980 ml. The readings for 0, 2.01, 4.96, 10.02, 14.98, and 20.13 ml of the standard copper solution were 40.7, 46.2, 55.5, 71.0, 86.2, and 100.0. Calculate by the graphical method the percentage of copper in the sample.

6–13.   A 1.64-g sample of wild duck feathers was ashed. The ash was dissolved in hydrochloric acid, diluted to 25 ml, and analyzed for lead by atomic absorption. The analysis showed a reading 1.32 times as great as that of a lead solution prepared by dissolving 0.3361 g of lead, diluting the solution to 100 ml, and then repeating a 10- to 250-ml dilution three times. What is the percentage of lead in the feathers?

6–14.   Compare atomic absorption on the basis of sensitivity, selectivity, and precision with (a) titrimetric, (b) radiochemical, and (c) colorimetric methods.

SELECTED REFERENCES

H. L. Kahn, *J. Chem. Educ.* **43**, A7 (1966); **43**, A103 (1966). A discussion of the instrumentation used for atomic absorption methods of analysis.

L. L. Lewis, *Anal, Chem.* **40,** 28A (No. 12) (1968). A perspective on the advantages and disadvantages of atomic absorption spectroscopy.

W. Slavin, *Atomic Absorption Spectroscopy,* Wiley (Interscience), New York, 1968. A practical treatment of the topic.

J. Topping, *Errors in Observation and Their Treatment,* Chapman & Hall, London, 1955. Straightforward information on least-squares calculations.

D. G. Peters, J. Hayes, and G. Hieftje, *Chemical Separations and Measurements,* W. B. Saunders Company, Philadelphia, 1974, Chapter 2.

6–5   SPECTROPHOTOMETRY OF A TWO-COMPONENT MIXTURE: DETERMINATION OF COBALT AND NICKEL AS EDTA COMPLEXES

Background

Spectrophotometric analysis of complex mixtures without prior separation is often possible. This experiment is designed to illustrate the application of the principles of spectrophotometry to a sample containing two absorbing substances. With mixtures of cobalt and nickel many complexes of the two metals have broad absorption spectra that overlap appreciably. Nevertheless, both can be determined by measuring the absorbances at two selected wavelengths if the molar absorption coefficient of each absorbing species at the two wavelengths is known. Because specificity is provided by the selection of two appropriate wavelengths, chemical separation is unnecessary. Least interference, highest sensitivity, and best precision are attained when wavelengths are chosen at which the absorbance differences between the species are maximized and spectral overlap is minimized (Figure 6-8). Both cobalt and nickel react with EDTA at pH values above 4 (Table 3-1). The resulting complexes are so

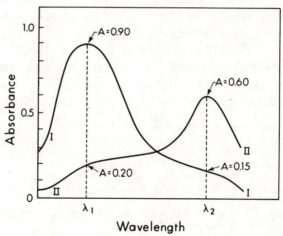

FIGURE 6–8. Spectra of two hypothetical compounds. The best wavelengths for analysis are λ_1 and λ_2.

stable that almost no other common ions interfere with development of the color, and a linear relation between absorbance and concentration of each complex exists. Although the complexes (and therefore the intensity of the colors) are stable, the rate of reaction of EDTA with nickel and especially with cobalt is slow. Solutions therefore have to be warmed long enough after addition of EDTA to ensure complete reaction.

The molar absorption coefficients of these complexes in the visible region are small (less than 100), so the procedure described is applicable only to relatively high concentrations of these metals.

Procedure (median time 4.7 hr)

Preparation of Standard Solutions of Cobalt and Nickel

Accurately weigh into a 100-ml volumetric flask enough cobalt metal powder to prepare 100 ml of approximately 0.05 M solution. Add 15 ml of dilute HNO_3 and place on a hot plate until dissolution is complete. Do not stopper the flask! Neutralize with dilute NaOH solution to the first permanent precipitate of cobalt hydroxide. Add a few drops of acetic acid to clear the solution.[1] Dilute to volume. Similarly, prepare 100 ml of 0.05 M nickel solution.[2,3]

Analysis Procedure

If the sample is a solid, weigh accurately an amount corresponding to 0.1 to 0.2 g each of cobalt and nickel and transfer to a 100-ml volumetric flask. The dissolution procedure depends on the composition of the sample. If it is a mixture of oxides, add 8 ml of concentrated HCl, dissolve on a hot plate, and neutralize the solution as directed above. (If the original sample is a liquid, omit the dissolution step.) Add acetic acid to clear the solution, dilute to volume, and mix.

Pipet 20-ml aliquots into two 50-ml volumetric flasks. Pipet 20-ml aliquots of cobalt standard and of nickel standard into four more 50-ml volumetric flasks (two for cobalt, two for nickel). Prepare 100 ml of buffer, 1 M in NH_4Cl and 1 M in NH_3. Add 10 ml of this buffer

---

[1] Sometimes a small amount of flocculent precipitate will remain. This is probably a fatty acid (less than 0.1%) added in preparation of the metal powder to aid in nucleation of the metal particles. Filtration usually is not required.

[2] If small amounts of insoluble residue remain after the dissolution of nickel powder, obtain another sample of metallic nickel. The residue is probably NiO, which does not dissolve readily in HNO_3.

[3] If insufficient NaOH is added, the acid remaining may exceed the capacity of the ammonia buffer to be added later.

to each volumetric flask, 1.9 g of the disodium salt of EDTA (0.005 mole) to the samples, and 0.8 g (0.002 mole) to each of the cobalt and nickel standards. The EDTA is most readily added with the aid of a powder funnel. Remove the stoppers and warm the flasks on a hot plate for about 2 hr to complete formation of the complex. Complexation is indicated by a marked color change. Cool and dilute to volume.[4] Prepare a blank containing 3 ml of buffer and 0.4 g of disodium EDTA in 25 ml of solution.[5]

Determine the absorbance at various wavelengths in the visible region for one cobalt and one nickel standard solution, taking readings at 20-nm intervals from 350 to 650 nm. Plot the cobalt and the nickel spectra on a graph, with absorbance on the vertical and wavelength on the horizontal axis. From the curves, choose two wavelengths for analysis and measure the absorbance of each of the standard and sample solutions at these two wavelengths. Make at least five sets of readings of the solutions at each wavelength. After each set place the blank solution in the sample compartment and readjust the absorbance to read zero.

Report the percentage of cobalt and of nickel in the sample.

Calculations

Spectrophotometric analysis of a mixture is based on the principle that each substance absorbs light independently of the others. Thus the observed absorbance at any wavelength is the sum of the individual absorbances of the species present in the mixture at that wavelength:

$$A_1 = a_{Co_1} bC_{Co} + a_{Ni_1} bC_{Ni} \qquad (6\text{--}16)$$

and

$$A_2 + a_{Co_2} bC_{Co} + a_{Ni_2} bC_{Ni} \qquad (6\text{--}17)$$

where A_1 and A_2 are the observed absorbances of the sample solution at the two wavelengths, Subscripts 1 and 2 refer to the first and second wavelengths chosen, a_{Co} and a_{Ni} are molar absorption coefficients, b is the path length, and C_{Co} and C_{Ni} are molar concentrations of the two metals in the sample. The molar absorption coefficients are calculated from the absorbance readings of the standards. By inclusion of the weights of cobalt and nickel taken, the absorbance readings of the standards, and

---

[4] Some white solid may appear upon cooling. This is probably excess EDTA, which should cause no difficulty. Allow any solid to settle before removing samples for spectral measurements.
[5] The blank need not be prepared in a volumetric flask.

the final volumes of solution, Equations (6-16) and (6-17) can be rearranged to

$$\% \, Co = (A_1 A_{Ni_2} - A_2 A_{Ni_1}) \, (g \, Co) \, (100)/[(A_{Ni_2} A_{Co_1} -$$

$$A_{Ni_1} A_{Co_2}) \, (g \, sample)] \qquad (6-18)$$

and

$$\% \, Ni = (A_1 A_{Co_2} - A_2 A_{Co_1}) \, (g \, Ni) \, (100)/[A_{Co_2} A_{Ni_1} -$$

$$A_{Co_1} A_{Ni_2}) \, (g \, sample)] \qquad (6-19)$$

where the symbol A_{Ni_2} is the absorbance of the standard nickel solution at the second wavelength, and so on. Note that b does not appear in either (6-18) or (6-19) because it is the same for both standards and samples and so cancels.

PROBLEMS

6-15. A 1.596-g sample was analyzed for cobalt and nickel by the procedure of this experiment. The weight of pure cobalt metal taken was 0.3247 g and of pure nickel metal 0.3014 g. The average absorbances recorded for the sample, cobalt standard, and nickel standard were 0.621, 0.538, and 0.097 at wavelength λ_1 and 0.644, 0.118, and 0.496 at wavelength λ_2. What are the percentages of cobalt and nickel in the sample?

6-16.† Chromium and manganese in steel may be determined by dissolution and then oxidation to chromate and permanganate, followed by spectrophotometric measurement at 450 and 525 nm. A solution containing 12.0 μg of manganese per milliliter gave a percent-transmittance reading of 76.1 at 450 nm and 32.0 at 525 nm; another solution containing 201 μg of chromium per milliliter gave readings of 40.0 and 72% at 450 and 525 nm. What was the chromium and manganese content of a solution that, under the same conditions, gave percent-transmittance readings of 36.1 and 45.0 at 450 and 525 nm?

6-17. Can the total amount of cobalt and nickel in a sample be determined by a direct EDTA titration?

6-18.† An acid HA has a dissociation constant of 1 × 10^{-4}. The undissociated species has molar absorption coefficients of 30 and 0 at 500 and 650 nm, and the anion has molar absorption coefficients of 0 and 200 at 500 and 650 nm. Calculate and plot the relation between the absorbance of a solution of HA in: (a) a 1-cm cell at 650 nm and its concentration over the range 0.1 to 0.001 M in pure water; (b) a solution buffered at pH 2; (c) a solution buffered at pH 4. Carry out a similar set of calculations and plots for the absorbance at 500 nm.

6–6   INFRARED SPECTROPHOTOMETRY AND COMPUTER ANALYSIS: DETERMINATION OF FOUR SUBSTITUTED BENZENES FROM SPECTRAL DATA

Background

In Experiment 6-5 a two-component mixture was resolved by making measurements at two wavelengths and setting up and solving two equations. A chemical separation was unnecessary. With sufficient data, spectrophotometric techniques of this type can be used to analyze mixtures of several components. Thus, one could in principle analyze three-, six-, or ten-component mixtures by making measurements at three, six, or ten wavelengths and solving three, six, or ten simultaneous equations. As the number of components increases, however, the quality of the instrumentation becomes more critical, and the arithmetic more tedious. In this experiment the tedium of the calculations required for the solving of four simultaneous equations is relieved through the use of a computer. Also, the application of infrared spectrophotometry to the resolution of a complex mixture is illustrated.

For this experiment sets of spectra run on a high-resolution infrared spectrophotometer are supplied. The samples consist of spectra of mixtures of p-dichlorobenzene, p-dibromobenzene, p-bromomethyl-benzene, and p-xylene that have been dissolved in a decane–bromoform

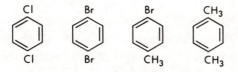

solvent. The infrared spectra of the four substituted benzenes are nearly identical except in the wavelength region of 12.2 to 12.7 μm (Figure 6-9). In this region infrared absorption by these compounds is due to symmetric, out-of-plane deformation vibrations of adjacent hydrogen

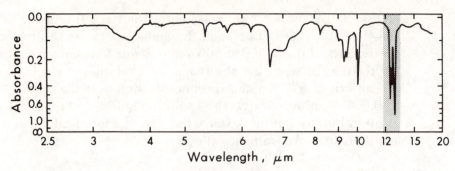

FIGURE 6–9. Infrared spectrum of a mixture of *p*-xylene, *p*-dichlorobenzene, and *p*-dibromobenzene in the ratio 3:1:2. The spectra of these compounds are the same except for the shaded region.

atoms on the ring. These vibrations are at right angles to the plane of the ring and give rise to intense absorption when the hydrogen atoms are vibrating in phase.

Mathematical Approach to Calculations

If values for several unknown quantities are to be determined, a number of independent equations equal to or exceeding the number of unknowns must be available. In this experiment, four quantities, the concentrations of four solutes, are sought. This evaluation requires four equations involving each of the four concentrations. The equations are obtained from the spectra, one for each of the four wavelengths at which absorbances are measured:

$$A_1 = a_{1w}c_w + a_{1x}c_x + a_{1y}c_y + a_{1z}c_z \tag{6–20}$$

$$A_2 = a_{2w}c_w + a_{2x}c_x + a_{2y}c_y + a_{2z}c_z \tag{6–21}$$

$$A_3 = a_{3w}c_w + a_{3x}c_x + a_{3y}c_y + a_{3z}c_z \tag{6–22}$$

$$A_4 = a_{4w}c_w + a_{4x}c_x + a_{4y}c_y + a_{4z}c_z \tag{6–23}$$

where A_1 is the absorbance of the sample at the first wavelength, a_{1w} to a_{1z} are the specific absorption coefficients of the four individual components of the mixture at this wavelength, and c_w to c_z are the concentrations of each of the four components in the sample. Each equation is simply a statement that the total absorbance is equal to the sum of the absorbances of the four components at a single wavelength.[1]

---

[1] The value of b is not included in the above equations because, as in Equations (6-18) and (6-19), it is constant for both standards and samples, and so cancels.

Four equations in four unknowns can be solved directly by combining equations to eliminate unknowns until one equation in one unknown is obtained. The procedure becomes unwieldy, however, when the number of unknowns exceeds three. Several systematic schemes are available to simplify the calculations. We use one called the Gaussian elimination procedure, which is particularly convenient for computer solution.

In Gauss's method a set of N equations in N unknowns is reduced to an equivalent triangular set, that is, a set having identical solution values, by the following procedure.

1. Equation (6-20) is divided by the coefficient a_{1w} of the first unknown c_w to give

$$cw + \frac{a_{1x}}{a_{1w}} c_x + \frac{a_{1y}}{a_{1w}} c_y + \frac{a_{1z}}{a_{1w}} c_z = \frac{A_1}{a_{1w}} \qquad (6\text{–}24)$$

Next, Equation (6-24) is multiplied by the coefficient of c_w in (6-21) a_{2w}, and the result subtracted from (6-21). This operation eliminates c_w from (6-21). Equation (6-24) then is multiplied by the coefficient of c_w in (6-22) a_{3w}, and the resulting equation is subtracted from (6-22) to eliminate c_w from (6-22). Similarly, c_w is eliminated from (6-23). The term c_w now has been eliminated from all but the initial equation; the first two equations in the new set are

$$a_{1w}c_w + a_{1x}c_x + a_{1y}c_y + a_{1z}c_z = A_1 \qquad (6\text{–}25)$$

and

$$\left(a_{2x} - \frac{a_{2w}a_{1x}}{a_{1w}} \right) c_x + \left(a_{2y} - \frac{a_{2w}a_{1y}}{a_{1w}} \right) c_y$$

$$+ \left(a_{2z} - \frac{a_{2w}a_{1z}}{a_{1w}} \right) c_z = A_2 - \frac{a_{2w}A_1}{a_{1w}} \qquad (6\text{–}26)$$

In this operation, (6-20) is called the *pivot* equation because it is used to eliminate an unknown from each of the equations that follows it.

2. In the same way, c_x is removed from the last two equations in the new set by using (6-26) as the pivot equation and repeating the procedure of Step 1.

3. The newly obtained expression for (6-22) from Step 2 is now used as the pivot equation, and c_y is eliminated from the fourth equation,

which is left with c_z as the only unknown. A set of four equations has now been obtained of the form

$$a\ c_w + b\ c_x + c\ c_y + d\ c_z = A_1 \qquad (6-27)$$

$$b'\ c_x + c'\ c_y + d'\ c_z = A'_2 \qquad (6-28)$$

$$c''\ c_y + d''\ c_z = A'_3 \qquad (6-29)$$

$$d'''\ c_z = A'_4 \qquad (6-30)$$

This set is called the *equivalent triangular* set.

4.   Equation (6-30) gives directly the value for c_z. This value is substituted in (6-29) to give c_y, and so on until all four concentration values have been obtained.

The original set of equations can be solved conveniently by matrix algebra. The initial matrix can be reduced efficiently with a computer, and the subsequent back substitution also is readily accomplished. For those interested, a general discussion of the Gaussian elimination method can be found in an appropriate mathematics reference text. The computer program employed is given in Figure 6-10.

Measurement of Absorbances from Recorded Spectra

In addition to the mathematical problems associated with quantitative multicomponent analysis, several points associated with infrared spectra should be considered. A problem in the extraction of reliable absorbance information from recorder tracings is deciding on the most accurate positions for the base lines and the absorbances at the absorption maxima. The position of the base line in a recorded spectrum is generally subject to error because of noise. It may also change with wavelength and therefore may not be parallel to the edge of the chart. Another effect of noise is that the absorbance at the absorption maximum often does not coincide with maximum excursion of the pen. With noise of short period the maximum absorption is likely to be less than the maximum pen excursion; with noise of longer period it may be either greater or less. When absorbances are being read from a recorded spectrum, the maximum of each peak should be drawn as though noise were absent. Figure 9-1 shows peaks without noise. Peak maxima may be distorted also by interferences; in the sample spectrum in Figure 6-11, the maximum for xylene may be distorted by the small absorption peak of p-bromomethylbenzene at about 12.7 μm.

```
C
C
C          PROGRAM TO CALCULATE  CONCENTRATIONS IN 4-COMPONENT  IR EXPERIMENT
C
C          UNIVERSITY OF ALBERTA CHEMISTRY DEPARTMENT
C
C
C              CREATE AUGMENTED MATRIX, ABSORB
C
           DIMENSION NAME(4),UNKN(3),ROOM(2),CMPD(3,4),CONC(4),ABSORB(4,5)
           READ(5,100)((CMPD(L,J),I=1,3),CONC(J),(ABSORB(I,J),I=1,4),J=1,4)
           READ(5,101)NAME,UNKN,ROOM,(ABSORB(I,5),I=1,4)
           WRITE(6,102)NAME,ROOM
           WRITE(6,103)(NAME,(CMPD(L,J),L=1,3),CONC(J),(ABSORB(I,J),I=1,4),
          .J=1,4)
           DO 1 J=1,4
           DO 1 I=1,4
 1         ABSORB(I,J)=ABSORB(I,J)/CONC(J)
           CALL SOLVE(ABSORB,CONC)
           WRITE(6,104)((CMPD(L,J),I=1,3),CONC(J),J=1,4)
           STOP
 100       FORMAT(16X,3A4,1X,F7.3,F8.3,2X,F8.3,2X,F8.3,2X,F8.3)
 101       FORMAT(7A4,1X,A4,A3,F8.3,2X,F8.3,2X,F8.3,2X,F8.3)
 102       FORMAT('1',4A4,15X,'ROOM # ',2A4///'0COMPUTER CALCULATION OF CONCEN
          .TRATIONS IN 4-COMPONENT IR EXPERIMENT.')
 103       FORMAT('0YOUR INPUT DATA WAS AS FOLLOWS:'//4(' ',4A4,5X,3A4,5X,
          .5F8.3/))
 104       FORMAT('0ON THE BASIS OF THIS DATA,THE CONCENTRATIONS IN WEIGHT PE
          .RCENT OF THE 4 COMPONENTS OF THE UNKNOWN ARE: '//4(' ',5X,3A4,
          .5X,F8.3/))
           END
C
C
C              THIS SUBROUTINE SOLVES FOUR SIMULTANEOUS EQUATIONS BY
C              GAUSSIAN ELIMINATION.
C
C
           SUBROUTINE SOLVE(A,X)
C
           REAL*4 A(4,5), X(4)
C
C              CREATE AN EQUIVALENT TRIANGULAR MATRIX.
           DO 20 I = 1, 3
           I1 = I + 1
           DO 10 J = I1, 4
           QUOT = A(J,I) / A(I,I)
           DO 10 K = I1, 5
 10        A(J,K) = A(J,K) - QUOT * A(I,K)
           DO 20 J = I1, 4
 20        A(J,I) = 0.0
C              BACK SUBSTITUTION TO DETERMINE X.
           X(4) = A(4,5) / A(4,4)
           DO 40 I = 1, 3
           SUM = 0.0
           IM = 4 - I
           I1 = IM + 1
           DO 30 J = I1, 4
 30        SUM = SUM + A(IM,J) * X(J)
 40        X(IM) = (A(IM,5) - SUM) / A(IM,IM)
           RETURN
           END
```

FIGURE 6–10. A fortran computer program for calculating concentrations of four unknowns from absorbance readings at four wavelengths. The program in use may include checks of the student cards for proper punching, and also print error messages if wrong columns or unreasonable numbers have been used. These portions are omitted here.

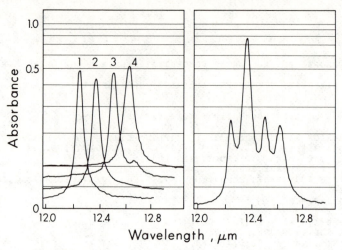

FIGURE 6–11. Infrared spectra of each of the four standards (left) and of a mixture of the four (right). Wavelength region shown corresponds to shaded portion of Figure 6–9.

Procedure (median time 2.8 hr)

Measurement Procedure

Obtain prerecorded infrared spectra of a sample mixture and of the four pure (standard) components. Bearing in mind the preceding discussion on the influence of noise on maxima and base lines, read the absorbance and base line of each standard at the four wavelengths corresponding to each of the four peak maxima. Subtract the base-line value from the absorbance reading in each case. Similarly, read the absorbance and base line of the sample spectra at each of the four wavelengths. Sixteen absorbance values for the standard curves and four for the mixture must be obtained.

Punching Absorbance Data onto Cards

Obtain a small pack of blank (unpunched) IBM cards. Punch five cards, one for each standard spectrum and one for the sample mixture, as follows.

1. **Columns 1 to 16.** Your initials and last name. If more than 16 letters, abbreviate.
2. **Columns 17 to 28.** Name of compound or, on the sample card, code number of sample. Abbreviate names of compounds if necessary. Code number for the sample must include the identifying letter as well as the digits, without spaces or characters between.

3.   **Columns 30 to 36.**   Concentration of standard solution or, on sample card, laboratory room number. Concentration values may be obtained from information supplied on the standard spectra. The units for the sample concentrations calculated by the program will be the same as for the standards. For proper operation of the Fortran program of Figure 6-10, all decimal points must be punched. On the sample card, indicate your laboratory room by the letter R followed immediately by the appropriate digits (no spaces between).

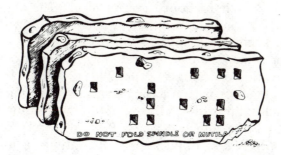

4.   **Columns 37 to 44, 47 to 54, 57 to 64, and 67 to 74.**   Absorbance values at the four wavelengths. Punch the absorbance readings in order of increasing wavelength. Each number must include a decimal point, and only digits and decimal points should be punched.

Check the printing at the top of each card to ensure that the punching is correct, and then arrange the five cards in the order *p*-dichlorobenzene, *p*-dibromobenzene, *p*-bromomethylbenzene, *p*-xylene, and sample. Submit the data cards as directed by the laboratory instructor. If an output message indicates an error in punching, correct the error and resubmit. Even if there is no error message, check the output to make certain it is reasonable; the computer accepts input data as punched and does not check for erroneous absorbances. Computers are machines without intelligence (garbage in, garbage out).

When a satisfactory output is obtained, report the results, with proper attention to significant figures.

Operating Instructions for IBM 29 Keypunch

1.   Turn on keypunch (red switch near right knee, marked "power on").
2.   Insert a few unpunched cards into the card hopper (right top) with the printed sides facing the keyboard. Bring the follower slide forward against the cards.

3. Make sure the toggle switches at the top of the keyboard are in the "up" position (except the "on/clear," which has a spring return).
4. Near and below the drum, visible through the window in the upper center of the keypunch, is a small, gray, plastic, V-shaped switch. The right side of the V should be pressed down.
5. Press the "rel" (card release) button on the keyboard once. A card should move down from the hopper.
6. Repeat Step 5.
7. Keypunch the data; the red index of numbers just below the drum indicates the column about to be punched. The numeric key, like the shift key on a typewriter, when held down punches the upper-case symbol.
8. If an error is made, press the "rel" key and start over with a new card. Correction of errors is not feasible.
9. To duplicate all or part of a card, put it in the hopper with a blank card behind it. Press the "rel" key until the already punched card is in position to be punched again. Press the "rel" key once more. Press the "dup" key; duplication will continue as long as it is held down.
10. After checking the cards (printing across top corresponds to punching), turn off keypunch.

PROBLEMS

6–19. In the plots of absorbance against wavelength in Figure 6-11, why are the absorbance scales logarithmic and inverted relative to those in Figure 6-8?

6–20. Estimate the uncertainty in an absorbance reading taken from Figure 6-11 for absorbance readings of 1.0, 0.7, 0.4, and 0.1.

SELECTED REFERENCE

D. A. Skoog and D. M. West, *Principles of Instrumental Analysis,* Holt, Rinehart and Winston, New York, 1971, Chapter 6. Introduction to infrared spectroscopy.

6–7 PHOTOMETRIC TITRATIONS: DETERMINATION OF IRON(III) WITH EDTA

Background

The concept of using low-precision measurements to obtain precise results for titrations is discussed in detail in Section 8-1. In the present

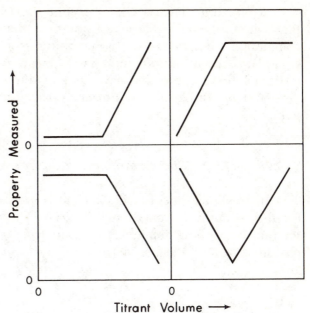

FIGURE 6–12. Examples of changes in a measured property as a function of amount of titrant added.

experiment light-absorption measurements further illustrate this concept. Conditions are adjusted so that the relation between volume of titrant and absorbance is linear before and after the end point; this linearity permits end-point selection from the intersection of two straight lines (Figure 6-12). Since measurements must be made at a number of points during an analysis, the reaction must be reasonably fast for the method to be practical.

An important advantage of this type of end-point determination is that measurements at the equivalence point are unnecessary, and accordingly special care in this region is not required. This advantage is illustrated in Figure 6-13, where readings of a changing property have been plotted against volume of titrant. Straight lines drawn through the points before and after the equivalence point are extended until they intersect to give an end point.

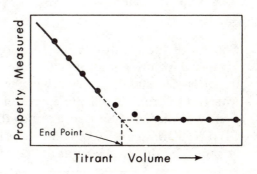

FIGURE 6–13. Selection of an end point by extrapolation of linear portions of a plot.

A reaction that has a small equilibrium constant is not complete in the region of the equivalence point. Excess titrant may drive it to completion, however, so those points somewhat distant from the equivalence point best determine its location. Thus the method is not restricted to reactions that have large equilibrium constants, but may be used also for those that are not analytically complete at the equivalence point.

Salicylic Acid as a Photometric Indicator for Iron(III)

In this experiment, salicylic acid is used as a photometric indicator for the titration of iron(III) with EDTA. The carboxylic acid portion of the molecule is a moderately strong acid (pK_1 = 2.97), while the hydroxyl proton is a very weak acid (pK_2 = 13.4). Several transition-metal ions

form chelate complexes with salicylate anion. For iron(III) the final complex has three salicylate anions ($Sal^=$) attached to each iron(III) ion. The log of the overall equilibrium constant for the formation of the complex is 34.5. In acid solution, competition for the phenolic (ROH) oxygen by hydrogen ion shifts the equilibrium to the point where EDTA can compete successfully with salicylate for iron(III), even though the log of the formation constant for the iron(III)– EDTA complex is only 25.1. The net reaction for the titration is

$$FeSal_3{}^{3-} + H_3Y^- + 3H^+ \underset{}{\overset{pH\ 2.2}{\rightleftharpoons}} FeY^- + 3H_2Sal \qquad (6\text{--}31)$$
$$\text{(blue-} \qquad\qquad\qquad\qquad\qquad \text{(faint}$$
$$\text{violet)} \qquad\qquad\qquad\qquad\qquad \text{yellow)}$$

Control of the pH during the titration is important. If the pH is too low, extensive protonation of salicylate will occur, and formation of the iron(III)–salicylate complex will be repressed; if the pH is too high, iron(III) hydroxide will precipitate. In this experiment the pH is adjusted first with hydrochloric acid and then a buffer of chloroacetic acid, $CH_2ClCOOH$ (pK = 2.3), and its sodium salt is added to maintain the pH at about 2.2 during the titration.

The end point is located by measurement of light absorption as a function of added reagent. Because of the high molar absorption coefficient at the absorption maximum, and the large formation constant for the iron(III)–salicylate complex, absorbance readings at that wavelength remain high until near the end point. Just before the end point the absorbance readings fall rapidly; after it they level out at a low value. The EDTA solution is standardized against pure calcium carbonate by a method similar to that in Experiment 3-3.

Procedure

Standardization of EDTA

Prepare an approximately 0.015 M EDTA solution by adding 4 g of the disodium salt of EDTA ($Na_2H_2Y \cdot 2H_2O$) and 3 ml of 6 M NH_3 to about 750 ml of water.

Weigh into a 100-ml beaker, to the nearest 0.1 mg, approximately 0.5 g of dry $CaCO_3$. Add 10 ml of water and then dissolve by rapid, dropwise addition of 11 ml of 6 M HCl. When dissolution is complete, transfer the solution quantitatively to a 100-ml volumetric flask and fill to the mark. Pipet 10-ml aliquots of this standard calcium solution and 10-ml portions of 0.02% $MgCl_2$ solution into 200-ml conical flasks. Immediately prior to each titration add about 6 ml of 6 M NH_3 solution and 4 to 5 drops of Calmagite indicator solution. Titrate with EDTA solution until the indicator changes from red to pure sky blue with no tint of red. Run a blank titration on a solution containing 1 ml of 6 M HCl and an accurately pipetted 10-ml portion of the $MgCl_2$ solution. Add NH_3 and indicator and then titrate as above. Subtract the volume required for the blank titration from the volume required for the sample to obtain the net volume of EDTA solution.

Preparation of Sample

Accurately weigh 0.6 g of sample into a 100-ml volumetric flask,[1] add 12 ml of concentrated HCl, and warm on a hot plate (in a fume hood) until dissolution is complete.[2] After dissolution neutralize with 6 M NaOH solution until $Fe(OH)_3$ just begins to form as a permanent

---

[1] Exercise care when transferring the sample to the volumetric flask, for the finely powdered material is easily lost. Use a funnel in the neck of the flask and rinse the funnel with a portion of the HCl. Dissolution requires about 2 hr.

[2] A small amount of white insoluble residue is silica and may be ignored. Any dark particles probably contain magnetite, Fe_3O_4, which is slow to dissolve in HCl. If the amount is small relative to the sample taken, it also may be ignored.

precipitate. Carefully clear the solution by adding a drop or two of $6\,M$ HCl, warming the solution if necessary. Cool and dilute to volume.

Selection of Wavelength for Titration

A suitable wavelength for the titration is selected from spectra of typical solutions before and after the end point. Obtain the spectra as follows. Pipet a 10-ml aliquot of the sample into a 50-ml volumetric flask, dilute to volume, and mix. Pipet a 10-ml aliquot of this solution into a 250-ml volumetric flask.[3] Add 2 ml of 6% salicylic acid[4] in methanol and 1.0 ml of chloroacetate buffer. Dilute to volume. Determine the approximate absorption spectrum for this solution with a spectro-photometer. (Operating instructions for the Bausch and Lomb Spectronic 20 spectrophotometer are given in Section 6-3.) Now add EDTA solution to a portion of the iron(III)–salicylate solution until the violet color disappears, and then 1 to 2 ml more. Determine the absorption spectrum for this solution. Plot the two absorption spectra on a single graph. The optimum wavelength for carrying out the photometric titration is the one at which the difference between the absorbances is greatest.

Titration of Sample

Pipet 10-ml aliquots of the original iron solution into 250-ml beakers, dilute to 50 ml with distilled water, and add 1 ml of 6% salicylic acid in methanol and 5 ml of chloroacetate buffer. Add standard EDTA solution from a buret until the color begins to fade. Then remove a portion of the solution from the beaker and measure the absorbance, which should be in the range 0.5 to 0.8. Return the solution to the beaker. Add 0.5-ml portions of EDTA, reading the absorbance after each addition until three or four readings have been taken before, and three or four beyond, the end point. Plot the absorbance against the volume of EDTA added. Determine the end point from the intersection of the linear portions.

Report the percentage of iron in the sample.

PROBLEMS

6–21.   A 0.5467-g sample of pure calcium carbonate was dissolved and diluted to 100 ml. A 9.993-ml aliquot of this solution was

---

[3] This serial dilution gives a final concentration equal to 0.008 of the original.

[4] A large excess of salicylic acid over iron ensures that all the iron will be in the form of $Fe(Sal)_3^{3-}$

titrated with 38.74 ml of EDTA for a standardization. A blank containing 10 ml of a 0.02% magnesium chloride solution required 1.42 ml of the EDTA solution. What was the molarity of the EDTA?

6–22.   A 9.993-ml aliquot of a 100-ml solution prepared by dissolution of 0.5844 g of iron ore required 48.24 ml of 0.01427 *M* EDTA solution. What was the percentage of iron in the ore?

6–23.   Sketch the shape of the photometric titration curve expected if the molar absorption coefficient of the iron(III)–salicylate complex were 10 times less than the value determined above.

6–24.   Sketch the shape of the photometric titration curves expected for the reaction

$$A + B \rightleftarrows C + D$$

if B is the titrant and at the wavelength used radiation is absorbed by: (a) A only; (b) C only; (c) A and C equally.

6–25.†   Calculate the equilibrium constant for the reaction shown in Equation 6-31. (*Hint:* Write the equilibrium expressions for the acid dissociations of EDTA and salicylic acid and for the formation of complexes of iron with EDTA and salicylate. Then combine them to give the equilibrium-constant expression.)

SELECTED REFERENCES

J. B. Headridge, *Photometric Titrations,* Pergamon Press, Elmsford, N.Y., 1961.
P. B. Sweetser and C. E. Bricker, *Anal. Chem.* 25, 253 (1953).
D. G. Peters, J. Hayes, and G. Hieftje, *Chemical Separations and Measurements,* W. B. Saunders Company, Philadelphia, 1974, Chapter 19.

Chapter 7

NONAQUEOUS ACID–BASE TITRATIONS

7–1 GENERAL BACKGROUND

Water is usually the preferred reaction medium because it is inexpensive, is easily obtained in pure form, and readily dissolves a wide variety of substances. Sometimes, however, an equilibrium may be unfavorable or a substance insoluble. In these cases another solvent, either inorganic or organic, may be used. Most inorganic liquids other than water, such as hydrazine, ammonia, or sulfur dioxide, have low boiling points and are toxic; but many organic solvents, such as methanol, acetone, or toluene, have a convenient liquid range at atmospheric pressure, are relatively unreactive, and are reasonably easy to purify. Many ketones, alcohols, and chlorinated hydrocarbons are prepared routinely in tank-car lots for use as solvents in a variety of chemical processes, from paint and plastics manufacturing to dry cleaning. We consider here systems in which the appropriate choice of an organic solvent makes possible analyses that cannot be carried out in water. To understand how such analyses become feasible, we first consider the behavior of acids in water and in glacial (100%) acetic acid.

When HCl is dissolved in water, the reaction may be written

$$HCl + H_2O \rightleftarrows H_3O^+ + Cl^- \tag{7-1}$$

The equilibrium for this reaction lies far to the right; in fact, the definition of a strong acid in water requires that ionization to yield H_3O^+ and the anion of the acid be essentially complete. Thus HCl and $HClO_4$ are both strong acids in water. If we change from water to a solvent that is a weaker proton acceptor, the reaction in (7-1) does not proceed so completely, and the position of equilibrium depends on the relative base

158

strengths of the solvent SH and the acid anion. Thus the equilibrium positions for the reactions

$$HCl + SH \rightleftarrows SH_2^+ + Cl^- \qquad (7–2)$$

and

$$HClO_4 + SH \rightleftarrows SH_2^+ + ClO_4^- \qquad (7–3)$$

depend on the basicity of the solvent. When the solvent is glacial acetic acid, it reacts only slightly with HCl but significantly with $HClO_4$. In acetic acid, therefore, $HClO_4$ is a strong acid and HCl is a weak acid. In practice, $HClO_4$ is one of the strongest acids available and is used to titrate a wide variety of bases in acetic acid.

Water is sufficiently basic that it reacts with many acids to form the hydronium ion, H_3O^+:

$$H_2O + HA \rightleftarrows H_3O^+ + A^- \qquad (7–4)$$

This tendency to convert strong acids to H_3O^+ is called the *leveling effect* of water. The hydronium ion is the strongest acid capable of existence in water, because any acid stronger than H_3O^+ reacts with water to produce hydronium ions. A similar leveling effect occurs in any solvent that is more basic than the acid being added. A solvent can level bases to the same strength in a similar way; the strongest base capable of existence in water is the hydroxide ion, because any base stronger than OH^- reacts with water to produce hydroxide ions.

7–2   NONAQUEOUS PHOTOMETRIC TITRATION: DETERMINATION OF OXINE WITH *p*-TOLUENESULFONIC ACID

Background

The use of a spectrophotometer to locate the end point in a titration (photometric titration) is discussed in Experiment 6-7. In the present experiment this method of end-point detection is used to determine the amount of a weak base, 8-hydroxyquinoline, by titration with a strong acid, *p*-toluenesulfonic acid, in the organic solvent acetonitrile. Additionally, a weight titration technique, in which the weight rather than the volume of added titrant is measured, is employed here because it has several advantages when only small quantities of sample are available and when organic solvents are involved. These advantages are discussed in the next section.

8-Hydroxyquinoline (oxine) is so weak a base in water ($pK_b = 9.8$) that it could not be titrated even if it were soluble enough to attain a useful concentration. In contrast, a solvent such as acetonitrile, CH_3CN, provides a medium in which oxine is readily soluble and can be titrated with an acid.

Acetonitrile is the solvent of choice for the determination of oxine in this experiment because it dissolves oxine readily and has little basic character. A solution of *p*-toluenesulfonic acid (*p*-TSA) is used as the titrant because it is a moderately strong acid, is stable in acetonitrile, and is available in highly pure form and so can be weighed directly as a primary standard. Solutions of perchloric acid and other strong acids in acetonitrile catalyze polymerization of the solvent.

Gravimetric Titrations

A gravimetric titration is one in which the weight rather than the volume of titrant is measured. The advantages are: (1) weight can be determined more accurately and precisely than volume; (2) burets, pipets, and volumetric flasks are not required; (3) calibration of volumetric equipment and the necessity of removing traces of grease for perfect drainage are eliminated; (4) small volumes of titrant and sample may be used, an advantage when reagents are expensive or when only small amounts of sample are available; and (5) errors in volume measurement due to expansion or contraction of solvent with temperature are avoided.

Despite these advantages, weight titrations have not been popular, principally because they may require many weighings during a titration, a tedious process with old-style balances. But with the advent of top-loading balances capable of giving weights quickly and accurately to within a milligram, weight titrations are now more feasible, and under certain conditions more desirable, than volumetric methods. This experiment illustrates two situations for which a weight titration is recommended. The first is where an organic liquid is used as solvent for the acid titrant. Acetonitrile has a cubic coefficient of thermal expansion almost five times that of water, and the change in volume is about 1 ppt/°C. (The thermal expansion of most organic liquids is in this range.) Thus small changes in temperature affect the volume of titrant appreciably. Also, the volatility of acetonitrile (bp 82°C) causes loss of solvent from an open container at an appreciable rate; the resulting change in titrant concentration may be considerable. The second situation is where it is desirable for convenience and greater precision to perform a photometric titration directly in a spectrophotometer cell without removal of the cell from the instrument. In these instances small volumes of titrant must be delivered while the relative accuracy and precision of conventional titrations is retained.

In the procedure used here the titrant is delivered from an ordinary

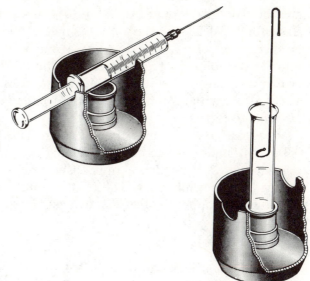

FIGURE 7–1. Support for syringe and titration cell during weighing.

10-ml syringe with a stainless-steel needle. The syringe serves as a convenient vessel for both weighing and delivering the titrant, and thus losses from evaporation are minimal. After each delivery of titrant the syringe is weighed to the nearest milligram on a top-loading balance. In this way a titration requiring, say, 2 g of *p*-TSA solution to reach the end point can be measured to better than 1 ppt. To accommodate somewhat larger volumes of sample, a test tube 25 by 150 mm is used as a cell. A stirrer designed to fit into the cell without obstructing the light path during readings expedites mixing and prevents loss of sample that would occur if the stirrer had to be removed for absorbance measurements (Figure 7-1).

Photometric end-point detection is used because oxine is too weak a base to give a sharp visual color change with an indicator. A stronger acid-solvent combination such as perchloric acid in glacial acetic acid might permit direct titration. Perchloric acid in acetic acid as solvent is widely used in commercial laboratories, but is somewhat inconvenient to prepare and standardize and is moderately hazardous if not handled properly. For these reasons, *p*-TSA in acetonitrile is used in this experiment.

Procedure (median time 2.9 hr)

Preparation of Sample and Titrant

Prepare a standard solution of *p*-toluenesulfonic acid monohydrate (*p*-TSA, formula wt 190.22) in acetonitrile by weighing to the nearest 0.1

mg about 0.5 g of pure p-TSA[1] into a small beaker. Weigh a clean, dry, glass-stoppered 50-ml flask to the nearest milligram on a top-loading balance.[2] Dissolve the p-TSA in a small amount of acetonitrile and transfer the solution quantitatively to the flask. Rinse the contents of the beaker into the flask with several small portions of acetonitrile. Dilute to about 50 ml with acetonitrile, stopper well, mix, and weigh the flask plus contents to the nearest milligram.[3]

Prepare a solution of the unknown sample as follows. Weigh about 0.3 g of sample to the nearest 0.1 mg and transfer it to a clean, dry, glass-stoppered 100-ml flask previously weighed on a top-loading balance to the nearest milligram. Dilute the solution to about 100 ml with acetonitrile, stopper the flask, and weigh again.

Titration Procedure

Obtain a spectrophotometer, a 10- to 15-ml cell (verify that the cell compartment on the instrument will accommodate the cell provided), a stirrer, and a 10-ml syringe.[4] (A Bausch and Lomb Spectronic 20 spectrophotometer with a sample compartment accommodating a 1-in. test tube is convenient. Operating instructions are given in Experiment 6-3.) Set the wavelength control to 540 nm.

Weigh the cell (Figure 7-1) on a top-loading balance to the nearest milligram, add 10 to 15 g of sample solution, and weigh again. Set the spectrophotometer to zero on the percent-transmittance scale. Insert the cell and add 10 drops of a 0.1% solution of 4-phenylazodiphenylamine indicator in acetonitrile. Insert a stirrer,[5] mix, cover the cell, and set the instrument to zero absorbance (100% transmittance).

---

[1] Reagent-grade p-TSA, obtained as the monohydrate, typically assays 99.8% or higher. It is stable and nonhygroscopic. If the purity of a particular lot of the acid is unknown, or in doubt, test it by titration of a pure sample of 8-hydroxyquinoline (best prepared by sublimation of a few tenths of a gram under vacuum), using the procedure of this experiment.

[2] The concentration of titrant is defined as grams of p-TSA per gram of solution. The flask must be dry, because excess water reduces the strength of the acid, and because acetonitrile or water initially in the flask is not included as part of the solution weight.

[3] Stock solutions of p-TSA and oxine must be kept tightly stoppered to prevent evaporation of the solvent. Once a portion of oxine solution has been weighed into the spectrophotometer cell, however, evaporation of solvent does not affect the amount of oxine present. The situation is analogous to pipetting an aliquot of sample into a flask for titration. Changes in total volume of solution after pipetting, either from evaporation or from addition of solvent when washing down the sides of the flask, do not affect the amount of sample being titrated.

[4] The syringe may be fitted either with a 22-gauge stainless-steel needle or with a glass or polyethylene tip having a fine orifice at the end. A thin film of silicone grease on the plunger allows control of the titrant addition and prevents titrant loss along the ground-glass surface. The syringe and cell are supported during weighing by a cradle constructed from a 16-oz polyethylene bottle, the upper part of which is cut at the shoulder to fit snugly into the bottom half (Figure 7-1).

[5] The stirrer, conveniently a polyethylene rod or a 16-gauge copper wire sealed in polyethylene tubing (Figure 7-1), must be positioned in the cell so that it does not obstruct the light path during measurements.

Draw 5 to 10 ml of *p*-TSA titrant solution into the syringe. Weigh the syringe to the nearest milligram, record the weight,[6] and add 0.4 to 0.5 g of titrant solution to the cell. Retract the plunger slightly after each portion of titrant is added, to minimize evaporation of the drop that would otherwise remain on the delivery tip. Stir the cell solution, cover the cell, and record the absorbance. Record the weight of the syringe, add another portion of titrant, and continue as before. When the absorbance reaches 0.08 to 0.1, take readings after each 1 or 2 drops of titrant up to an absorbance of about 0.45, when the titration may be terminated.

To determine the weight of titrant required to reach the end point, plot weight of titrant against absorbance and draw straight lines through the points before and after the break point (Figure 7-2). A major source of error in this experiment is insufficient care in drawing the lines used to select the end-point weight. The line after the break point should be drawn through the steepest portion of the curve.

Calculations

From the weight of *p*-TSA and the weight of solvent plus *p*-TSA, calculate the concentration of *p*-TSA titrant in grams of *p*-TSA per gram of solution. From the weights of titrant (obtained from the graph) and

---

[6] If the top-loading balance has a built-in taring device, set the optical scale to its upper limit, usually 10 g, at this point. (A check of the optical-scale sensitivity of the balance at this time is suggested.) In this way the total weight of titrant added at each step, obtained by subtracting the values recorded from the initial reading, is determined more readily. Some balances have a built-in complementary scale that allows the weight of titrant *removed* from the syringe to be read directly from the scale.

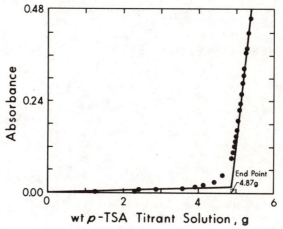

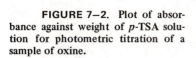
FIGURE 7–2. Plot of absorbance against weight of *p*-TSA solution for photometric titration of a sample of oxine.

sample, calculate the concentration of oxine in the sample. The percentage of oxine is obtained by

$$\% \text{ oxine} = \frac{(g\ p\text{-TSA/g titrant}) (\text{wt titrant}) (\text{formula wt oxine})}{(\text{mol wt } p\text{-TSA}) (\text{wt sample soln}) (g\ \text{sample/g soln})} (100)$$

The formula weight of oxine is 145.16.

PROBLEMS

7–1.   A 0.3101-g sample of impure oxine was dissolved in acetonitrile to give a total weight of solution of 77.938 g. An 11.200-g portion was titrated by the procedure of this experiment. The p-TSA titrant solution was prepared by dissolving 0.9623 g of p-TSA in acetonitrile to give 39.035 g of solution. The following data were obtained.

Wt Titrant	Absorbance
0.483	0.007
0.967	0.010
1.293	0.012
1.543	0.014
1.852	0.018
2.110	0.021
2.361	0.093
2.506	0.201
2.592	0.264
2.731	0.362
2.814	0.415
2.947	0.498

What was the percentage of oxine in the sample?

SELECTED REFERENCES

Gravimetric Titrimetry, Technical Information Bulletin No. 1014, Mettler Instrument Corp., Princeton, N.J., 1967.

J. B. Headridge, *Photometric Titrations,* Pergamon Press, Elmsford, N.Y., 1961.

D. J. Pietrzyk and J. Belisle, *Anal. Chem.* **38,** 969 (1966). Study of the acidity of aromatic sulfonic acids and their use as titrants in nonaqueous solvents. The nitro derivatives are the strongest of the benzene sulfonic acids, but somewhat expensive and not readily obtained in pure form.

H. A. Laitinen and W. E. Harris, *Chemical Analysis,* 2nd ed., McGraw-Hill, N.Y., 1974, Chapter 4.

D. G. Peters, J. Hayes, and G. Hieftje, *Chemical Separations and Measurements,* W. B. Saunders Company, Philadelphia, 1974, Chapter 5.

E. A. Butler and E. H. Swift, *J. Chem. Educ.* **49,** 425 (1972). Discussion of advantages of weight titrations in undergraduate teaching laboratories.

Chapter 8

ELECTROCHEMICAL METHODS OF ANALYSIS

8–1. GENERAL BACKGROUND

In the volumetric experiments in Chapters 3 and 4, equivalence points are determined by visual indicators. Experiments 6-7 and 7-2 illustrate that changes in a property of a system, for example, absorbance, may be followed during titration to indicate when a stoichiometric quantity of titrant has been added. Sometimes, however, suitable indicators are not available, or the solution to be analyzed is colored or turbid. In these situations the titration often can be followed by monitoring the concentration of a species electrometrically. For instance, the pH during an acid–base titration can be followed with a pH meter and a reference electrode–pH electrode pair, or the potential during an oxidation-reduction titration monitored with a potentiometer or volt-meter and a reference electrode–platinum electrode pair. A plot of pH or potential against volume of titrant added then permits precise location of the end point.

The concentration of an acid or base cannot be determined accurately by direct measurement with a pH meter. A single pH measurement normally has an accuracy and precision of only a few hundredths of a pH unit, which for a 0.1 M solution of a strong acid is a relative uncertainty of over 2% per 0.1 pH unit. Also, direct pH measurement does not give the total acid content of a solution when a buffer is present. Therefore, for accurate determinations of acidity at appreciable concentrations, the pH is measured with a pH meter as titrant is added, and the amount of acid or base present determined from a plot of pH against volume of titrant. During a titration exact values of pH are not required; only the change in pH with volume of titrant is important. For example, the concentration of an acetic acid solution can be measured with a precision of a few tenths of a percent by titration with standard base, even though pH measurements are recorded only to the nearest pH unit.

Concentrations of many species can be monitored with a variety of methods other than pH or potential measurement, such as measurement of current, light absorption, conductivity, temperature, or radioactivity. All can be put to analytical use.

Potentiometric Titrations

Potentiometric titrations provide an example of the use of low-precision measurements to obtain precise results. In Experiments 8-2 and 8-3 the relation between the volume of titrant added and the potential of the indicator electrode is not a linear function, so location of the end point is somewhat less straightforward than where the observed quantity is directly proportional to the titrant volume, as in Experiments 6-7 and 7-2. Because the electrode potential of a reversible couple varies exponentially with concentration, changes are large near the equivalence point, where the concentration of material being titrated is small (Figure 8-1). The end point often can be found from a plot of electrode potential against volume of titrant. Sometimes a section of the curve near the end point is expanded and plotted as rate of change of voltage against volume (Figure 8-2). Here the end point is located at the maximum of the curve. Plots of this type, known as first-derivative plots, often give better estimates of an end point than conventional plots. This procedure is recommended for the potentiometric determination of cerium(IV) by titration with ferrocyanide in Experiment 8-3.

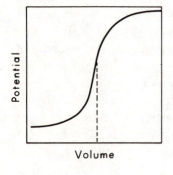

FIGURE 8–1. Plot of potential against volume of titrant in the equivalence-point region for a 1:1 reaction. Dashed line indicates end point.

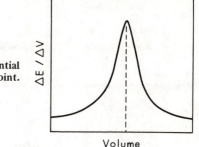

FIGURE 8–2. Plot of first derivative of potential against volume of titrant. Dashed line indicates end point.

Constant-Current Coulometry

Sometimes an analyst may wish to use a titrant that is either unstable in solution or difficult to prepare or store. Often the application of constant-current coulometry can be useful in these circumstances. In this technique the titrant is electrochemically generated directly in the solution in which it is to be used; the amount of the substance being analyzed for is found from the magnitude of the current and the time for which it flows. Because electronic instrumentation is now capable of providing specified currents constant to within 0.1%, and because the time of current passage can be accurately measured, results by this method can be highly accurate and precise. Preparation and standardization of a titrant solution is unnecessary, the question of titrant stability during storage is avoided, and any convenient type of end-point detection may be chosen.

The relation between current, time, and the number of equivalents of reagent generated is given by

$$\text{equivalents} = It/96,487 \qquad (8–1)$$

where I is the current in amperes, t the time in seconds, and 96,487 the number of coulombs (1 coulomb = 1 ampere of current flowing for 1 sec) required to generate one equivalent of a substance. This number is called the faraday.

Coulometry is not limited to generation of oxidizing or reducing agents. Hydroxide and hydrogen ions may be produced by electrolysis of water:

$$2H_2O + 2e^- \rightarrow H_2 + 2OH^-$$

$$2H_2O \rightarrow 4H^+ + O_2 + 4e^-$$

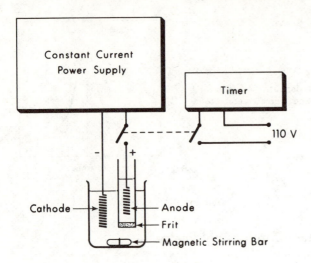

FIGURE 8–3. Schematic diagram of apparatus for constant-current coulometric analysis.

The titrant species is generated in the solution being analyzed; products formed at the other electrode are isolated from the solution so that they do not interfere. The second electrode is usually isolated in a separate compartment and electrical contact made through a fine glass frit. Figure 8-3 illustrates a coulometric system for generation of a reducing agent.

Another method of coulometric analysis, controlled-potential coulometry, involves generation of current at an electrode maintained at a constant potential. The potential is selected so that only the reaction of interest occurs at the electrode surface. The current therefore depends on the concentration of the substance being determined; as the concentration decreases, the current falls until both reach a negligible level. The total current is related to the amount of the species of interest by (8-1) and is measured with an electronic or chemical integrating device. This method is often more rapid and less subject to interferences than constant-current coulometry, but is not so accurate.

8–2. POTENTIOMETRIC TITRATION OF AN ACID MIXTURE

Background

In this experiment a mixture of two acids is titrated with standard base. Because the pH does not change with sufficient abruptness at either equivalence point to permit accurate determination of end points by visual indicators, a pH meter is used to measure the pH in the regions of the equivalence points. From a plot of pH against volume of base added, the amount of each acid present can be determined, and the dissociation

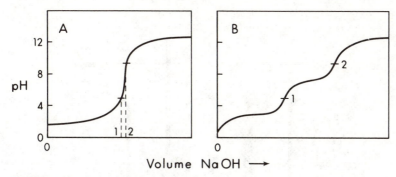

FIGURE 8–4. Plot of titrations with 0.1 M NaOH of sulfamic acid (left) and a mixture of two acids having pK's of about 3 and 7 (right).

constants of the acids often can be estimated. In the same way, mixtures of bases when titrated with standard acid produce a curve of the general shape of that shown in Figure 3-1; the size and location of the breaks depend on the K_b values of the two bases.

A mixture of two acids can be analyzed with a precision of a few parts per thousand if the dissociation constants differ by 10^3 to 10^4 or more and if the weak acid is not too weak. The molarity of the NaOH titrant is determined by titration of a standard acid. For greatest precision the standardization end point is taken at the same pH as the end point of the acid being titrated; this procedure corrects for the error caused by carbonate that may be present in the base. Because two end points at differing pH values are obtained in the titration of the mixture of acids, the calculation of two different concentrations for the same NaOH titrant may be required (Figure 8-4).

pH Meters and pH Measurement

The electrometric measurement of pH involves measurement of the potential between two electrodes, one a reference and the other an indicator electrode, responsive to changes in pH (Figure 8-5). The most important indicator electrode for pH is the glass electrode, discussed below. The resistance of a glass electrode is high, on the order of 10^7 to 10^9 ohms. For this reason, and also because the current drawn must be kept low to avoid affecting the measurement, the meter used to measure the potential must be highly sensitive. (A current of 10^{-10} ampere causes a voltage drop across 10^8 ohms of 10 millivolts, which is equivalent to 0.2 pH unit.) The current flowing in the electrode circuit therefore must be kept to values so low that considerable amplification is necessary before the signal can be read from the meter. An amplifier this sensitive, however, normally varies somewhat in response (drifts) during operation.

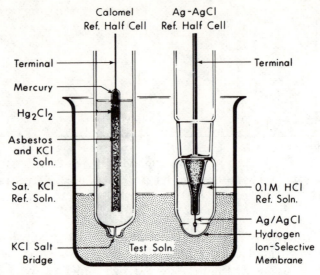

FIGURE 8–5. Schematic diagram of calomel reference and glass indicator electrodes for pH measurement.

To correct for drift in the meter, and also to compensate for variations in the electrical characteristics of individual glass electrodes, daily calibrations of meter and electrodes must be carried out on a solution of known pH before unknown solutions are measured. The potential registered by a pH meter also is affected by the temperature of the solution being measured, and so this factor, too, must be taken into account in accurate measurements.

The Glass Electrode

Probably more pH measurements are carried out each day then any other type of chemical measurement, most of them made potentiometrically with a glass indicator electrode. The properties of this electrode required over fifty years of study before its mode of operation was understood; and even today improved glass membranes are developed mostly by empirical methods.

A typical electrode consists of a thin glass membrane, generally bulb shaped, sealed onto the end of a glass tube. A constant pH is maintained on the inside of the bulb by filling it with an internal buffer solution. Contact between the buffer and the lead to the meter is made through an internal reference electrode (generally a calomel or silver–silver chloride reference). The buffer solution and internal reference electrode usually are sealed into a closed unit for stability and convenience in handling (Figure 8-5).

Measurement of the pH of a solution involves immersion of the glass bulb, along with an external reference electrode, in the solution and observation of the potential developed between the two electrodes. The resulting cell may be written

| internal reference electrode; buffer solution | ‖ | membrane | ‖ | solution under measurement | | external reference electrode |

The vertical double lines indicate the junctions of the surfaces of the membrane with solution, and the vertical single line the junction between the external reference electrode and the solution under measurement. The potential at each of these junctions remains essentially constant when the test solution is changed, except at the interface of the membrane with the solution under measurement. At this junction the potential varies with the hydrogen ion activity of the solution under test according to the relation

$$E = E' - \frac{2.303 \, RT}{F} \log \frac{1}{a_{H^+}} \qquad (8\text{–}2)$$

In (8-2) E' depends on the nature of the reference electrode used, the history of the membrane, and several other factors. Because some of these factors vary from day to day, a pH meter and an indicator electrode–reference electrode pair are standardized before each set of measurements by placing the electrodes in a solution of known pH and adjusting the pH meter to read that pH value. In effect, a value for E' in (8-2) is determined by this procedure.

The factor $(2.303 \, RT)/F$ in (8-2) equals 0.0591 V at 25°C. Since this factor varies with temperature, most pH meters include a control that can be set to the temperature of the test solution to provide the necessary correction.

The surface of a glass membrane responsive to pH consists of a hydrated layer of glass about 0.1 μm thick. (The total thickness of the membrane is approximately 50 μm.) In this surface layer, lithium or sodium ions present at lattice sites in the glass can exchange with hydrogen ions in solution by an ion-exchange type of equilibrium. The extent to which exchange occurs depends on the hydrogen ion concentration in the solution. This cation equilibration produces an electromotive force across the membrane that is a function of the pH of the solution. As mentioned, some current must flow across the membrane if this potential is to be measured. The mechanism of current flow through the major part of the glass can be thought of as a slight shifting of cations, such as sodium or lithium, already present in the lattice sites in the glass. Although each ion moves only a short distance, the net result is a transfer of charge across the membrane.

The two surfaces of a glass membrane do not behave exactly alike, but differ in response, depending on the history of the individual surface. These variations in response produce a small but observable potential difference across the membrane even when the solutions on each side are identical. This difference, called the asymmetry potential, is generally on the order of a few millivolts. The magnitude of this potential varies from one electrode to the next, and from day to day for a given electrode. It is corrected for at the time an electrode is standardized with a reference solution, and because the rate of change is slow, rechecking during a pH measurement is usually unnecessary. After the first standardization in a series of measurements, most adjustments needed during restandardization compensate for drift in the pH-meter amplifier.

To be useful, the glass electrode must be reasonably sturdy. Membrane thicknesses are chosen to compromise between mechanical strength and chemical resistance on the one hand and accurate and rapid pH response on the other. The membrane is susceptible to physical breakage and to chemical attack, especially from hydrofluoric acid and highly alkaline solutions. The normal life of an electrode is approximately 1 to 2 years, but is shortened if the membrane is frequently allowed to dry.

According to Equation (8-2), the response of a glass electrode is 0.0591 V per tenfold change in hydrogen ion activity at 25°C. Present commercial glass electrodes follow this relation closely over the pH range of about 1 to 13. Above or below these values response tends to fall off somewhat, and some error is introduced.

The affinity of sites in the glass matrix for hydrogen ions over other cations is a major factor in the usefulness of the glass pH electrode. Since other cations do not compete appreciably with hydrogen ions under most conditions, interferences are few. In early electrodes containing sodium glasses this selectivity began to break down at sodium hydroxide concentrations of about 10^{-4} M (pH 10), and error was introduced because the ratio of sodium ions to hydrogen ions became sufficiently large for sodium to interfere. This error amounted to a pH unit or more in 0.1 M solutions of sodium hydroxide. Glasses fabricated with lithium instead of sodium are not subject to sodium ion error at sodium hydroxide concentrations as great as 1 M (pH 14); such glasses are used in almost all present-day pH electrodes. New glasses have also been developed that are more responsive to cations other than hydrogen ions, and glass electrodes are now available that show selectivity for sodium, for lithium, and for silver ions.

It can be seen that the accuracy of pH measurements with a pH meter and a glass pH electrode depends in large part on the accuracy of the pH standard solution and the care with which the standardization operations are carried out. In fact, pH is best defined in terms of values assigned to certain solutions of substances of known stability and purity. Thus, the U.S. National Bureau of Standards has specified pH values for a series of reference solutions. Four of the most important ones are (all at 25°C): 0.05 M potassium tetroxalate, 1.679; 0.05 M potassium hydrogen phthalate, 4.008; 0.025 M potassium dihydrogen phosphate–0.025 M disodium hydrogen phosphate, 6.865; and 0.01 M borax, 9.180. Although these values are specified to the nearest 0.001 pH unit, they are considered to be accurate to only 0.006, and under laboratory conditions measurements accurate to 0.01 pH unit are of high quality.

Procedure (median time 3.5 hr)

Preparation of 0.1 M NaOH Solution

Gently boil about 800 ml of distilled water in a 1-liter flask for 4 or 5 min to remove dissolved carbon dioxide. Cover the flask with a watch glass and allow the water to cool to room temperature. Transfer about 500 ml to a clean polyethylene or borosilicate-glass bottle. Add sufficient NaOH to make an approximately 0.1 M solution and mix.[1] Keep tightly covered to minimize absorption of CO_2.

Standardization of NaOH Solution

Weigh, to the nearest 0.1 mg, samples of approximately 0.25 g of sulfamic acid into 250-ml beakers. Add 50 ml of boiled and cooled water to each. Insert glass and reference electrodes into the solution so that the bulb of the glass electrode and the junction at the tip of the reference electrode are immersed in the solution. Titrate with 0.1 M NaOH solution, recording the buret readings each 0.1 to 0.2 pH unit throughout the end-point region. Stir the solution thoroughly after each addition of base; a magnetic stirrer is convenient for this purpose. Plot the titration data directly on a graph during the titration to clarify the pH changes through the end-point region. Continue the titration until the plot begins to level off at about pH 10.

---

[1] Because reagent-grade NaOH may contain considerable Na_2CO_3, it is frequently dispensed as a saturated solution (about 50% NaOH by weight), in which the solubility of Na_2CO_3 is low. In this experiment small amounts of Na_2CO_3 in the NaOH titrant are compensated for in the standardization operations. Nevertheless, care should be taken that CO_2 absorption does not occur between the time of standardization and titration of the substances being determined.

Titration of Unknown Mixture of Acids

Weigh, to the nearest 0.1 mg, 0.5- to 0.6-g samples of the acid mixture into 250-ml beakers.[2] Dissolve the samples in water and titrate as for the standardization above.[3]

Calculate and report the milliequivalents of each acid in the mixture.

Procedure for Operation of pH Meter[4]

1.   Turn the operating switch to the standby (STDBY) position and plug the instrument into an electrical outlet. Allow about 5 min for warmup. In the standby position, power is provided to the instrument, but the lead to the glass electrode is disconnected and the amplifier input is shorted. In this position the electrodes may be transferred from one solution to another, rinsed, or handled without damage to the instrument, while the amplifier remains warmed up and ready for use. Keep the operation switch in the standby position between measurements.

2.   To calibrate the meter, immerse the electrodes in a reference solution of known pH. Set the TEMP control to read the temperature of the solution. Turn the operating switch to pH, and adjust the CALIB control to set the meter needle to the pH of the reference solution.

3.   Turn the operating switch to STDBY and remove the electrodes from the reference solution. Rinse the electrodes, wipe them with a tissue, and immerse them in the solution to be measured. Turn the operating switch to pH; the instrument will now indicate the pH of the solution. The switch may be left in the pH position during the course of the titration, as long as the electrodes are not removed from the solution.

4.   When finished with the last titration, turn the operating switch to STDBY and unplug the instrument power cord. Rinse the electrodes and immerse them in a beaker of distilled water for storage.

Calculations

Plot the data for each of the titrations and draw a smooth curve through the points. The end points for the acid mixture are best located

---

[2] These directions are for compounds with equivalent weights in the range of 150 to 200.

[3] If the pK values of the unknown acids are to be obtained, the pH meter and electrode must be calibrated against a reference solution of known pH.

[4] The procedure outlined is for a Corning Model 5 pH Meter, but applies in general to most meters in current teaching use.

by determining the inflection points in the curves. To do this, first draw the steepest tangent to the curve in the end-point region and then locate the points where the smoothed curve departs from the tangent. The end point is located midway between the two points of departure.

Sodium carbonate present in the NaOH titrant causes the molarity of the solution (relative to titration of an acid) to vary as a function of the strength of the acid. For example, if one titrates a solution of NaOH containing Na_2CO_3 with a strong acid such as HCl, $CO_3^=$ is converted to H_2CO_3 by the HCl, and the molarity of the base is determined by the total NaOH plus Na_2CO_3 present in the solution. On the other hand, if one titrates a weak acid with a pK of 1×10^{-6}, Na_2CO_3 in the NaOH solution is converted only to $NaHCO_3$, and a different molarity is obtained. Thus the molarity varies with pH. The best way to correct for carbonate error is to obtain from the standardization curves two NaOH molarities. The first is calculated from the volume of base required in the standardization to reach a pH identical to the pH at the first end point in the titration of the unknown mixture. The second is calculated similarly, from the pH corresponding to the second end point in the titration of the unknown mixture. Using the NaOH molarity calculated at the lower pH value, calculate the millimoles per gram of the stronger acid in the mixture. In a similar way, calculate the millimoles per gram of the weaker acid in the mixture, using the NaOH molarity calculated at the higher pH value. Assume both acids are monoprotic. From the titration curves, estimate the dissociation constants of the two acids.

Report the millimoles per gram of each acid in the unknown mixture.

PROBLEMS

8-1.   A mixture of acids required 22.43 and 38.37 ml of standard NaOH for titration to the first and second end points by the procedure of this experiment. If the molarities of the NaOH solution were 0.1058 and 0.1012 for titration to the first and second end points, how many equivalents of each acid were present in the sample?

8-2.   What additional information is needed to calculate the molecular weights of the two acids in Problem 8-1?

8-3.   How would the two molarities of the NaOH solution as determined by this experiment be affected if additional carbon dioxide were absorbed by the solution?

SELECTED REFERENCES

W. J. Blaedel and V. M. Meloche, *Elementary Quantitative Analysis,* 2nd ed., Harper and Row, New York, Chapter 18. Thorough introductory treatment of practical aspects of acid–base titrations.

H. A. Laitinen and W. E. Harris, *Chemical Analysis,* 2nd ed., McGraw-Hill, New York, 1974, Chapters 3, 6, and 13.

D. G. Peters, J. Hayes, and G. Hieftje, *Chemical Separations and Measurements,* W. B. Saunders Company, Philadelphia, 1974, Chapters 4 and 11.

8–3. POTENTIOMETRIC TITRATIONS: DETERMINATION OF CERIUM(IV) WITH POTASSIUM FERROCYANIDE

Background

In this experiment a sample containing cerium(IV) is dissolved in dilute sulfuric acid and the cerium(IV) concentration of the resulting solution determined by titration of an aliquot of standard ferrocyanide solution. The reaction is

$$Ce^{4+} + Fe(CN)_6{}^{4-} \rightleftarrows Fe(CN)_6{}^{3-} + Ce^{3+} \qquad (8-3)$$

The potential at an inert platinum electrode in the solution increases as the ratio of ferricyanide to ferrocyanide increases. After all the ferrocyanide has been oxidized, the potential changes rapidly to values readily calculated from the ratio of cerium(IV) to cerium(III).

Cerium(IV), a powerful oxidant in a solution of strong acid, is a useful analytical titrant. The electrode potential of the cerium(IV)–(III) couple depends on the acid concentration of the solution, the extent of hydrolysis of cerium(IV), and the presence of complexing anions. For prevention of hydrolysis and subsequent precipitation of cerium(IV) hydroxide, cerium(IV) salts are first dissolved completely in concentrated acid and then diluted slowly with stirring. If concentrated acid is not used, a clear solution may be obtained initially from which cerium(IV) hydroxide will precipitate later. A properly prepared solution of cerium(IV) in sulfuric acid is stable indefinitely and is as convenient a titrant as permanganate or dichromate.

A vacuum-tube voltmeter (VTVM), a potentiometer, or a pH meter (a highly sensitive voltmeter) is suitable for measuring the potential of the platinum indicator electrode. A second electrode, one that does not change in potential as the concentrations of the oxidizing and reducing agents vary, is required to complete the measuring system. This is called a *reference electrode.* The most common reference electrode is the saturated calomel electrode (SCE), which has an electrode potential against the standard hydrogen electrode of 0.244 V;[1] however, any electrode whose potential does not change during the titration can be used. The reference electrode in this experiment is a silver wire coated with silver chloride and immersed in a solution of sodium chloride. The potential of the electrode is determined by the silver(I) activity, which in turn depends on the solubility product of silver chloride and the chloride ion activity. The Nernst equation for the silver(I)–silver couple is

$$E = E^0 {}_{Ag^+, Ag} - \frac{RT}{nF} \ln \frac{1}{[Ag^+]} \tag{8–4}$$

For an aqueous solution at 25°C, a chloride ion concentration of 0.1 M, and activity coefficients of unity, the equation may be written

$$E = E^0 {}_{Ag^+, Ag} - 0.0591 \log \frac{0.1}{K_{sp}} \tag{8–5}$$

where K_{sp} is the solubility product for silver chloride, 1.8×10^{-10}, and 0.0591 is the value of 2.303 RT/F at 25°C for a one-electron process. The factor 2.303 is for conversion from base-e to base-10 logarithms. Substituting numerical values for E^0 and K_{sp} gives

$$E_{Ag^+, Ag} = 0.799 - 0.0591 \log (5.6 \times 10^8) = 0.282 \text{ V} \tag{8–6}$$

A convenient way to arrange for electrical contact between the reference-electrode and titration solutions is through a ground-glass joint (Figure 8-6). Mixing of the solutions is generally undesirable because it

---

[1] The potential of the standard saturated calomel half-cell mentioned here includes a junction potential between the half-cell and a buffer solution of 0.1 ionic strength. This is more applicable to general experimental work than the value omitting junction potentials, 0.2412 V. For discussion of this point see *Reference Electrodes*, D. J. G. Ives and G. J. Janz, Eds., Academic Press, New York, 1961, Chapter 3.

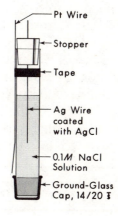

FIGURE 8–6. Assembly for reference electrode. A platinum wire taped to the assembly as shown serves as a convenient indicator electrode.

Pt Wire
Stopper
Tape
Ag Wire coated with AgCl
0.1M NaCl Solution
Ground-Glass Cap, 14/20 ⚛

results in loss of part of the solution being titrated and because both ferrocyanide and ferricyanide form insoluble precipitates with silver that affect the potential of the reference electrode. Despite the appreciable electrical resistance of the ground-glass contact between the two solutions, accurate potential readings can be obtained because a vacuum-tube voltmeter requires only minute current for measurement.

An application of the cerium(IV)–ferrocyanide reaction is the indirect determination of reducing sugars. An excess of ferricyanide is added to a sugar sample in alkaline solution (pH 10), and the mixture allowed to stand for a period of time that depends on the sugar. Ferrocyanide is formed in an amount equivalent to the sugar present. After reaction the solution is acidified and the ferrocyanide titrated with standard cerium(IV) solution.

Procedure (median time 3.0 hr)

Preparation of Standard Ferrocyanide Solution

Weigh accurately 1.7 g of $K_4Fe(CN)_6 \cdot 3H_2O$ into a beaker. Dissolve in distilled water and transfer to a 100-ml volumetric flask. Add about 1 g of Na_2CO_3 and dilute to volume.[2]

Analytical Procedure

Since many cerium(IV) salts are hygroscopic, weighing by difference is recommended to minimize water absorption. Using a triple-beam balance, add approximately 5.0 g of the sample to a weighing bottle. Dry at least overnight in an oven at about 110°C. Cap the weighing bottle and allow to cool. Accurately weigh the bottle and sample. Transfer most of the sample to a small beaker. Recap and accurately reweigh the bottle and residual sample.

Add 2 to 3 ml of 18 *M* sulfuric acid (concentrated reagent) to the sample. (*Caution*: Concentrated sulfuric acid is extremely corrosive to bench tops, clothing, and people. Handle only in area provided. Also, the addition of water results in liberation of considerable heat. Watch for spattering. Safety glasses are essential.) Stir thoroughly into a smooth paste (about 2 min). Gradually add more acid to a total of about 10 ml and stir another 2 to 3 min. Add 5- to 10-ml portions of water, stirring for at least 2 min after each addition, until the sample is almost completely

---

² Sodium carbonate is added to stabilize the ferrocyanide, which decomposes slowly on standing in acid solution. The solution should be used within a few hours of preparation.

dissolved; from 60 to 100 ml of water will be required. Quantitatively transfer the solution to a 250-ml volumetric flask, dilute to volume, and shake vigorously for at least 5 min to complete dissolution. (A slight suspension of solid at this point can be ignored. Consult your laboratory instructor if in doubt.)

Prepare approximately 0.1 M NaCl for the reference electrode by weighing about 0.3 g of NaCl to the nearest milligram, dissolving in water, and diluting to volume in a 50-ml volumetric flask.

When ready to begin the titrations, obtain a vacuum-tube voltmeter[3] and reference-electrode assembly. Coat the silver wire to be used as a reference electrode with silver chloride by removing the ground-glass cap and immersing the assembly in 0.1 M HCl. Connect the silver wire to the positive terminal and the platinum wire to the negative terminal of a 1.5-V battery, and allow current to flow for about 30 sec. Disconnect the battery, rinse the assembly with distilled water, replace the cap, and partially fill the reference electrode with the previously prepared 0.1 M NaCl solution. Ensure that the ground-glass surfaces are completely wetted by the solution. Store with the end immersed in a beaker of distilled water, and rinse the outside with water before using.

Pipet 20 ml of the standard ferrocyanide solution into a 150-ml beaker and add 25 ml of water. Turn on the meter and allow it to warm up for a few minutes. With the leads shorted, zero the meter with the zero control. When ready to titrate, add about 5 ml of 3 M H_2SO_4 to the ferrocyanide solution and dip the reference electrode into the solution. Only the Pt wire and the tip (including the ground-glass junction) of the reference electrode should come in contact with the solution being titrated. Connect the silver wire in the reference solution to the negative (or reference electrode) terminal of the meter and the platinum wire to the positive (or indicator electrode) terminal. Set the meter switch to (−) or (+) volts or millivolts, whichever causes the meter to read on scale.[4] The sign of the switch position is the sign of the potential of the platinum wire in the test solution relative to the reference electrode. Select a voltage-range setting of 1 or 2 V.

Titrate with cerium(IV) solution in increments of about 3 ml for the first sample. It is worthwhile to plot the curve as the data are obtained. For succeeding samples only the values near the end point are important, so these titrations may be carried rapidly to within 1 to 2 ml of the end point before the potentials are recorded. For these samples, titrate dropwise near the end point. After each addition of titrant, stir the

---

[3] The directions in this experiment apply in substance to most vacuum-tube voltmeters or pH meters equipped with a millivolt readout.

[4] The (−) position may be required initially. After a small amount of titrant has been added, the potential will read zero. From this point on, the (+) position should be used, because the solution potential at this point has become positive with respect to the reference electrode. On vacuum-tube voltmeters the DC Volts position should be used.

solution with the reference electrode, read the buret, and then read the meter. To aid anticipation of the end point, arrange the buret so that the titrant enters near the platinum wire. Determine the end point by taking the first derivative of the curve of voltage against volume of titrant and plotting it against the volume of titrant (Figure 8-2).

Report the percentage of cerium in the sample.

Calculations

Analytical Calculations

Determine the end point for each titration by plotting the first derivative of the titration curve. The plot is made by graphing $\Delta E/\Delta Vol$ on the vertical axis against Vol on the horizontal axis. Here ΔE is the difference between two successive potential readings, and ΔVol the difference in the corresponding volume readings. Vol on the horizontal axis is taken to be the average of the two values used to obtain ΔVol. In Table 8-1 a plot of $\Delta E/\Delta Vol$ against Vol shows a peak at 10.13 ml, which is the end point.

In the calculation of the percentage of cerium(IV), the weight in grams of cerium(IV) per titration is obtained from the relation

$$g\ Ce = [ml\ K_4 Fe(CN)_6]\ [M\ K_4 Fe(CN)_6]\ (g\ Ce/mole)/(1000) \tag{8-7}$$

The weight of sample taken per titration is

$$wt\ sample = \frac{(ml\ Ce\ used\ in\ titration)\ (total\ wt\ sample)}{250\ ml} \tag{8-8}$$

TABLE 8-1. EXAMPLE OF TABULATION OF DATA FOR
FIRST-DERIVATIVE CURVE

Buret Reading, ml	Potential, E	$\Delta E/\Delta Vol$	Vol
10.10	1.45		
		0.05/0.02	10.11
10.12	1.50		
		0.20/0.02	10.13
10.14	1.70		
		0.10/0.04	10.16
10.18	1.80		

Supplementary Calculations

This experiment can provide additional useful information. The formal potential for the ferricyanide-ferrocyanide couple can be calculated from the titration data; also, the potential of the reference electrode can be obtained. Each should be reported relative to the standard hydrogen electrode. The *standard electrode potential* for an oxidation-reduction couple is defined as the potential of that couple relative to the hydrogen electrode, with all species present under standard conditions (25°C and unit activities). Standard potentials often are not useful in practical terms because of unknown activity coefficients, complexation, and other effects. In such cases formal potentials are frequently used. *Formal electrode potentials* are potentials measured under conditions where the ratio of the formal concentrations (formula weights per liter) of the oxidant and reductant is unity and where the analytical concentrations of other substances present are specified. Use of formal potentials allows the conditions under which a measurement is made to be defined and reproduced. Formal electrode potentials may vary significantly from standard electrode potentials. For example, the standard potential for the cerium(IV)–(III) couple is 1.61 *V*, whereas the formal potential in 1 *M* sulfuric acid is 1.44 *V*.

PROBLEMS

8–4.   A 5.220-g sample of an impure cerium(IV) salt was analyzed by the procedure of this experiment. The volume of cerium(IV) solution required to titrate a 19.95-ml aliquot of 0.0406 *M* ferrocyanide was 38.54 ml. Calculate the percentage of cerium(IV) in the sample.

8–5.   A cerium(IV) hydroxide sample was analyzed by the procedure of this experiment. The data were: weight cerium(IV) sample, 4.6741 g; volume ferrocyanide taken, 19.97 ml; weight potassium ferrocyanide trihydrate taken, 1.7284 g; and volume cerium(IV) solution required to reach the end point, 41.20 ml. Calculate the percentage of cerium(IV) in the sample.

8–6.   A 0.5678-g sample of a tin ore is treated to give a solution of tin(II). Titration of the reduced sample requires 44.44 ml of 0.1111 *M* cerium(IV) sulfate. What is the percentage of tin(IV) oxide in the sample?

8–7.   Calculate the electrode potential of a solution 90% to the equivalence point in a titration of 0.2 *M* tin(II) chloride with standard 0.1 *M* cerium(IV) in 1 *M* sulfuric acid.

8–8.   For a potentiometric titration of 50 ml of 0.1 M ferrocyanide with 0.1 M cerium(IV), calculate the potential of the indicator electrode against a standard hydrogen electrode (a) before any cerium(IV) is added, (b) at the midpoint, (c) at the equivalence point, and (d) 100% past the equivalence point.

8–9.   A copper rod is dipped into pure water. What is the electrode potential (relative to a standard hydrogen electrode) if 10 copper(II) ions are present per liter of water; 2 ions; 1 ion; no copper(II) ions?

8–10.†   What is the formal electrode potential of the ferricyanide-ferrocyanide couple in 1 M $HClO_4$ if K_a for $HFe(CN)_6{}^{3-}$ is 6 X 10^{-5}?

8–11.†   One ml of 0.1 M sodium benzoate was added to 50 ml of 0.1 M silver nitrate. The potential of a silver electrode in this solution was measured against a reference electrode and found to be +0.300 V. After addition of 80 ml of 0.1 M sodium benzoate the potential was +0.164 V. Calculate the solubility product of silver benzoate.

SELECTED REFERENCES

J. J. Lingane, *Electroanalytical Chemistry,* 2nd ed., Wiley, New York, 1958, p 129.

D. G. Peters, J. Hayes, and G. Hieftje, *Chemical Separations and Measurements,* W. B. Saunders Company, Philadelphia, 1974, Chapters 9 and 11.

G. F. Smith, *Cerate Oxidimetry,* 2nd ed., G. F. Smith Chemical Co., Columbus, Ohio, 1964.

R. B. Whitmoyer, *Anal. Chem.* 6, 268 (1934). Ferricyanide is used to oxidize sugars such as glucose by addition of excess reagent and back titration with standard cerium(IV) sulfate.

8–4. CONSTANT-CURRENT COULOMETRY: DETERMINATION OF CYCLOHEXENE WITH ELECTROCHEMICALLY GENERATED BROMINE

Background

In this experiment the amount of cyclohexene in a methanol solution is determined by reaction with bromine. The reaction is

The volatility of bromine makes preparation and storage of standard solutions difficult. These problems are avoided by electrochemical generation of bromine through electrolysis of potassium bromide at a known constant rate in a mixture of acetic acid, methanol, and water.

The equivalence point is located by measurement of the current flowing between two indicator electrodes across which a potential is applied. This method, sometimes called *amperometry,* is based on the requirement that, for a current to flow between a pair of electrodes, electroactive species capable of undergoing oxidation or reduction at the applied potential must be present in the solution. The magnitude of the current that flows depends on the nature of these electroactive species and on their concentrations. In this experiment a potential of 0.2 V is applied across the indicator electrodes. Prior to the equivalence point the possible reactions are the oxidation of bromide at one electrode and the reduction of hydrogen ion at the other. Because the potential required to produce bromine and hydrogen is considerably greater than 0.2 V, essentially no reaction occurs, and flow of current between the indicator electrodes is negligibly small. At the equivalence point the first excess of bromine in the solution makes it possible for the reaction at the cathode to be reduction of bromine rather than of hydrogen ion. Since a potential difference of 0.2 V between the two electrodes is more than sufficient to oxidize bromide at one and reduce bromine at the other, the equivalence point is marked by current flow between the indicator-electrode pair.

In practice it is convenient to generate bromine in the cell until a current of some convenient magnitude is reached in the indicating circuit, and then to add sample to the cell and generate bromine again until the same current is obtained. This method has two advantages: the end point is more easily reproduced; and any impurities in the supporting electrolyte that react with bromine are corrected for. In effect, a blank has been run. The relation between the current in the indicating circuit and time is shown schematically in Figure 8-7.

Under the conditions of this experiment, the addition of bromine to

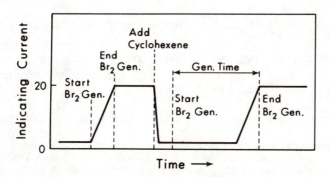

FIGURE 8–7. Schematic representation of change in current between indicator electrodes during blank and sample runs.

cyclohexene is slow unless a catalyst such as mercury(II) acetate is added. Different alkenes (compounds containing two aliphatic carbons connected by a double bond) undergo bromination at different rates. Two alkenes that react slowly with bromine are acrylic acid (CH_2=CHCOOH) and acrylonitrile (CH_2=CHCN).

Procedure (median time 3.0 hr)

Obtain an unknown sample of cyclohexene. Place in a dry 100-ml volumetric flask, dilute to volume with methanol, and mix well. Acquaint yourself with the experimental apparatus, identifying the source of constant current, the amperometric end-point detection system, and the cell components. Consult your laboratory instructor regarding any questions.

Add electrolyte solution, 0.15 M KBr in 60:26:14 (volume %) glacial acetic acid:methanol:water, to the cell until the generating electrodes are just covered. Add 0.1 g of mercury(II) acetate and a magnetic stirring bar. Adjust the stirring rate for maximum mixing without the bar striking the electrodes.

Set the voltage across the indicator-electrode pair to 0.2 V, and generate bromine at the anode until the current in the indicating circuit is 20 microamperes (μA). Generate bromine in small increments as the 20-μA level is approached, waiting after each increment until the level of the indicating current stabilizes. Pipet 5 ml of sample into the cell, replacing the stopper immediately after the addition. Generate bromine until the indicating system shows 20 μA, using small increments near the end point as before. Record the time required.[1] Pipet a second sample into the same solution and again generate bromine to the 20-μA end point. This operation may be repeated several times in the same electrolyte solution.

Calculate and report the milligrams of cyclohexene in the original sample.

Operating Instructions for Coulometric Titration Apparatus[2]

1. Connect the cathode of the generating circuit (Figure 8-8) to the negative terminal and the anode to the positive terminal of the coulometer.

---

[1] Some coulometers read directly in microequivalents of reagent generated rather than in time.

[2] The apparatus described here includes a Leeds and Northrup coulometer together with an amperometric end-point detection system consisting of a mercury cell as voltage source in series with a 1000-ohm potentiometer and a microammeter. Contact with the solution is made with two platinum wires (Figure 8-8).

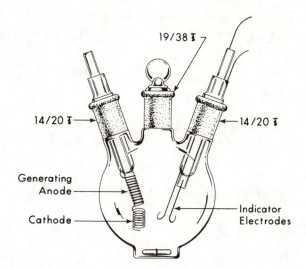

FIGURE   8–8. Arrangement of generating and indicator electrodes in cell for constant-current generation of bromine.

2. Turn on the Power switch of the coulometer. The Line Freq-Std Freq switch should be in the Line Freq position. (*Caution*: Do not touch the leads from the coulometer while the Output switch is on. Up to 300 V is supplied to these leads at open circuit.)
3. Set the Output, Milliamp control to 6.43 mA. Turn the reset knob to set the microequivalent readout to zero.
4. When ready to generate bromine, turn the Output switch on. Near the end point, turn the Output switch off and on as needed to generate small increments of bromine, analogous to dropwise addition of conventional titrant near the end point.
5. After completing a titration, record the microequivalents of bromine generated and reset the counter to zero. Another sample then may be added and the process repeated.
6. After completing a series of titrations, verify that the Output and Power switches are off. Disconnect the coulometer leads to the cell. Rinse the cell and electrodes thoroughly with water.

Calculations

The milligrams of cyclohexene present in the original 100-ml sample can be obtained from the relation

$$mg = (\mu eq) \left(\frac{0.08215}{2} \ mg/\mu eq \right) \left(\frac{100}{5} \right) \qquad (8\text{–}9)$$

The microequivalents of bromine generated are related to the current in milliamperes and to the time of generation in seconds by

$$\mu eq = \frac{(sec)\,(mA)\,(10^3)}{96{,}487} \tag{8-10}$$

The factor 2 is required in the denominator of (8-9) because two bromide ions must be oxidized for each molecule of cyclohexene.

PROBLEMS

8–12.   A 100-ml sample of a solution containing cyclohexene was analyzed by bromination of a 5-ml portion by the procedure of this experiment. If a current of 6.43 mA was required for 483.6 sec to reach the amperometric end point in the titration, how many milligrams of cyclohexene were present in the original 100-ml sample?

8–13.   Write the equations for the reactions taking place at the anode and cathode connected to the coulometer during the bromination of cyclohexene by the procedure of this experiment. Why is it necessary to isolate one of the generating electrodes in many coulometric reactions but not in this one?

8–14.   Write the equations for the reactions taking place at each of the indicator electrodes used to locate the amperometric end point (a) before the end point and (b) after the end point.

8–15.†   Devise a method for determining the equivalent weight of an alkene whose identity is unknown. Could the molecular weight also be determined?

8–16.   If an alkene that undergoes bromination relatively slowly were analyzed with the procedure and apparatus of this experiment, would the results likely be low or high? Why?

SELECTED REFERENCES

D. H. Evans, *J. Chem. Educ.* **45**, 88 (1968). Discussion of the method used in this experiment.

J. J. Lingane, *Electroanalytical Chemistry*, Interscience, New York, 1958, p 536.

D. G. Peters, J. Hayes, and G. Hieftje, *Chemical Separations and Measurements*, W. B. Saunders Company, Philadelphia, 1974, Chapter 12.

Chapter 9

RETRIEVAL OF INFORMATION AND TREATMENT
OF DATA

In this chapter the acquisition of quantitative information from strip-chart recorders is considered. Although this is only one of a variety of means by which data may be obtained from instruments, it is a major one, and it illustrates a useful and reliable approach. Another important method, reading meters, is employed in the use of many optical and electrochemical analytical instruments (see, for example, Experiments 6-3, 6-4, 6-5, 8-2, and 8-3). Still another, digital readout, is growing rapidly as advances in electronics provide sophisticated, inexpensive ways of presenting data as direct numbers. The advantages of digital readout, illustrated by a coulometric titration and a radiochemical measurement (Experiments 8-4 and 10-5), are that meter-reading errors are avoided and time is saved. It is important to keep in mind, however, that nonsignificant digits appearing in a digital readout often may give a false sense of precision and accuracy.

In addition to these methods of obtaining data from instruments, attention should be paid to the many readings that are taken in weighing and measuring volumes during both instrumental and noninstrumental analyses. For these readings the importance of techniques that minimize errors has been emphasized. Nevertheless, bias can creep into even such operations as reading a buret or a vernier scale. Section 9-1 discusses a statistical test for detection of such bias.

9–1. EVALUATION OF PREJUDICE IN DATA: CHI-SQUARE TEST

> Coincidences, in general, are great stumbling blocks in the way of that class of thinkers who have been educated to know nothing of the theory of probabilities—that theory to which the most glorious objects of human research are indebted for the most glorious of illustrations.
>
> *Edgar A. Poe*

187

Statistical quantities such as means, standard deviations, and confidence limits have sensible meaning only when applied to data free of determinate error. As pointed out in Section 2-1, however, determinate error may arise in any experimental situation. It is therefore important to ascertain whether bias is being introduced into experimental results by determinate factors. One objective test suitable for examining small numbers of observations for determinate error is the Q test. This test, described in Section 2-1, is recommended for the evaluation of data obtained in most chemical work.

When frequencies of occurrence of numbers can be expressed quantitatively in terms of probabilities, statistical tests may be applied. In this section such a test, the chi-square test, is introduced and its application to evaluation of the probability of prejudice is considered.

Chi-Square Test

A determinate factor can be introduced into experimental observations through prejudice that arises, for example, from a knowledge of preceding members of a set of replicates. In most chemical measurements containing four significant figures, the terminal digit should have a random distribution. However, in repetitive measurements of a quantity, say titrations of a series of aliquots, a tendency may creep in to have later buret readings coincide with early ones. Suppose that in a set of 47 observations the digits 0, 1, 2, . . ., 9 occurred at frequencies of 3, 7, 3, 5, 4, 2, 4, 8, 7, and 4. These might be the last digits in final buret readings or in analytical weighings. Is this the frequency distribution expected from chance, or does number prejudice (bias) exist? To evaluate the probability of prejudice, a chi-square value is calculated as in Table 9-1, where f is the observed frequency and F is the average, or expected, frequency. The average frequency is the sum of all the digits times the probability, or 47 $\times$ 0.1. Here the probability is 0.1 because the experimental observation can be any one of ten digits; it arises from a system that has nine degrees of freedom. The value of chi square for the above set is calculated by dividing the sum of $(F - f)^2$ by the average frequency. In this example the value of chi square is 37/4.7, or 7.9. The probability that any given chi-square value will be exceeded through pure chance can be determined by consulting a chi-square table for nine degrees of freedom (Table 9-2). More extensive tables can be found in handbooks of chemistry and in books on statistics. Interpolation of Table 9-2 indicates that the chi-square value 7.9 calculated from the data of Table 9-1 has a probability of about 0.55. When the probability is about 0.5, there is no evidence that number prejudice or other nonrandom factors have introduced a determinate bias into the observations. When the probability becomes appreciably greater

TABLE 9-1. DATA FOR CALCULATION OF CHI SQUARE

Terminal Digit	Frequency, f	$F - f$	$(F - f)^2$
0	3	1.7	3
1	7	−2.3	5
2	3	1.7	3
3	5	−0.3	0
4	4	0.7	1
5	2	2.7	7
6	4	0.7	1
7	8	−3.3	11
8	7	−2.3	5
9	4	0.7	1
	$\Sigma f = 47$		$\Sigma[(F - f)^2] = 37$
	$F = 4.7$		

or less than 0.5, it can be concluded with a high level of confidence that number prejudice exists.

The situation encountered most frequently in experimental work is prejudice for certain numbers and against others. For example, 0 and 5 typically appear with greater frequency than would be due purely to chance. In such cases chi-square values greater than 8.3, corresponding to probability values less than 0.5, are obtained. A strong suspicion of number prejudice is warranted for a small number of observations with probabilities in the range 0.2 to 0.05. A larger number of observations is needed to either confirm or discount the suspicion. At 0.05 the probability of prejudice is 95 out of 100, or 95%. At 0.001 the probability is 999 out of 1000—a high level of confidence.

Chi-square values indicating probabilities of appreciably more than 0.5 are rare. A chi-square value of unity would mean that all numbers were observed with equal frequency, a highly improbable condition. Very low chi-square values indicate that prejudice is being introduced by the observer, consciously or unconsciously, making number frequencies come out more even than they normally would as a result of chance. At a probability of 0.99 (chi square = 2.1) the chances are 99 out of 100 that the data are prejudiced.

PROBLEMS

9–1.   A student recorded the following frequencies for the last digit of the final buret readings for a series of titrations (last digit in parentheses): (0) 16, (1) 6, (2) 10, (3) 6, (4) 14, (5) 12, (6) 8, (7) 8, (8) 16, (9) 4. What is chi square and the probability of

TABLE 9-2. RELATION BETWEEN CHI SQUARE AND PROBABILITY FOR NINE DEGREES OF FREEDOM

Probability	0.999	0.99	0.90	0.50	0.20	0.10	0.05	0.01	0.001	0.0001
Chi Square	1.5	2.1	4.2	8.3	12.2	14.7	16.9	21.7	28	32
Likelihood of prejudice	Very high	High		Lowest			High			Exceedingly high

prejudice for this set of numbers? How would you interpret the results?

9–2.   One of a pair of dice was suspected of being tampered with. The following frequencies were observed for the numbers 1 through 6 for a series of rolls. Die 1: (1) 8, (2) 18, (3) 11, (4) 19, (5) 9, (6) 25. Die 2: (1) 11, (2) 14, (3) 12, (4) 19, (5) 19, (6) 15. Given the following table, calculate chi square and the probability of prejudice for each. What is the likelihood that one of the dice shows prejudice? How would you interpret the results?

Probability	0.99	0.90	0.50	0.20	0.10	0.05	0.01
Chi square	0.55	1.6	4.4	7.3	9.2	11.1	15.1

9–2. RETRIEVAL OF INFORMATION FROM STRIP-CHART RECORDERS: MEASUREMENT OF AREAS OF CHROMATOGRAPHIC PEAKS

Background

To obtain a permanent record of the results of an experimental measurement, scientists more and more are using recording instruments. Generally the measurement is converted to an electrical signal that is fed to a data collection device. This device is often a recorder having as output a strip of paper on which the measured quantity is traced as a function of some property such as time, reagent volume, or wavelength. The plot that results may be a peak, seen in spectrophotometry and chromatography, or an S-shaped curve, seen in titrations and some electrochemical methods such as polarography. From measurements of peak heights, widths, or areas, or from curve inflection points, both qualitative and quantitative information about a sample can be obtained. For example, in the spectrophotometric determination of iron (Experiment 6-3), the atomic-absorption determination of copper (Experiment 6-4), and the potentiometric titration of cerium(IV) (Experiment 8-3) the procedures could be modified so that absorbance or potential would be recorded directly on a chart rather than read from a meter.

Information obtained from recorder output must be related somehow to properties of interest in the sample. Quantitative information can be obtained in a variety of ways. A survey of practicing gas chromatographers in the United States (see reference at the end of this section) indicated that manual methods of peak evaluation were used for 66% of the measurements and disc integrators for 21%. The remaining 12% of the time a recorder was bypassed in favor of equipment such as a digital integrator. Of the manual methods, height, height–width, and related techniques accounted for 45% of the total; planimetry, 16%; and the cut-and-weigh technique, 6%.

Three experiments exemplify a range of techniques useful in retrieving information from recorder charts. Experiment 11-5 illustrates a quantitative determination of separated components in a sample by measurement of gas-chromatographic peak areas. Under the conditions of that experiment, the peak shapes are nearly ideal and the measurements relatively straightforward. Frequently, however, and especially in chromatographic work, peaks may be narrow and high, flat and broad, or small and asymmetric. Furthermore, particularly in measurements of radiant energy, the base line may be noisy. In Experiment 6-6 the infrared absorption peaks recorded on a logarithmic scale illustrate problems of overlap and instrument noise. In the present experiment three manual methods for evaluating Gaussian peaks of several shapes and sizes without noise (Figure 9-1) are compared.

Manual Integration Techniques

Height–Width. The area of a Gaussian peak can be obtained by measuring its height and width and then multiplying the product of these measurements by a proportionality factor. The value of the factor depends on the fraction of the height at which the peak width is measured.

The precision and accuracy of the measurement of peak heights and widths are highly dependent on technique. Four major operations, and

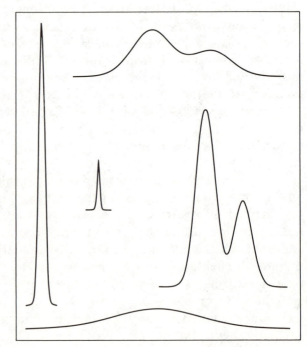

FIGURE 9–1. Some Gaussian chromatographic peaks without noise.

consequently four sources of indeterminate experimental error, are. involved: placement of the base line, measurement of the height from the base line, positioning of the measuring instrument at a predetermined intermediate height, and measurement of the width at that height. The standard deviation of the third operation, positioning, is about twice that of base-line placement, of height measurement, or of width measurement of a sharp peak. For flat peaks the width measurement is subject to large relative error. If all four errors are combined statistically for peaks of constant area but of varying height-to-width ratio, the relative error in peak area reaches a minimum at a height-to-width ratio of about 3 (Figure 9-2). As expected, for both sharp and flat peaks the relative error is larger than for more nearly optimum shapes. Also, for peaks of a given shape the relative error in height-width measurements decreases with increasing area as the square root of the area.

Although the width can be measured at any fraction of the height, it is slightly advantageous to measure the width of broad peaks at half height and the width of sharp peaks near the base line. When the areas of a full range of smooth Gaussian peaks are to be measured, the best single value to choose is measurement of peak widths at one quarter of the peak height. The peak area is then given by 0.753 × height × width. A template of the type shown in Figure 9-3 is convenient, although not essential, for locating the one-quarter position.

Height Alone. The evaluation of peaks by height measurement alone requires only two operations: placement of the base line and measurement of the height. The precision of this method is therefore inherently greater than for measurement of peak area—greater by an order of magnitude for some narrow peaks. Unfortunately, peak heights are sensitive to small instrumental and operational variations, so determinate rather than indeterminate errors limit the precision. This technique is of most value in carefully controlled routine gas chromatography. Since peak areas are much less sensitive to determinate variations, area measurements are preferred in nonroutine work.

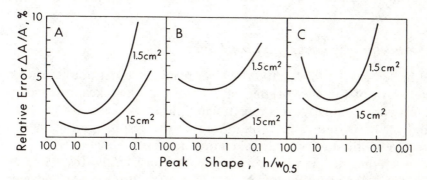

FIGURE 9–2. Relative error in measurement of area as a function of peak shape for peaks of 1.5- and 15-cm² area by (A) height–width, (B) planimeter, and (C) cut-and-weigh methods.

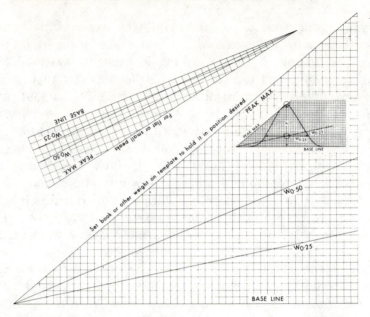

FIGURE 9–3. Transparent template for convenient location of quarter- or half-height position on peaks.

Perimeter Methods. The perimeter methods of planimetry and of cutting and weighing are appropriate for peaks of irregular shape as well as for Gaussian peaks.

When a planimeter is used, there is again an optimum peak shape for minimum error; the relative error, however, does not increase so drastically when peaks become flatter or sharper as it does in the height–width method (Figure 9-2B). For peaks of a given shape the relative error decreases with increasing size according to the three-quarter power of the area. The planimeter technique therefore is best suited to peaks of large area.

Variations in paper density make the cut-and-weigh method highly vulnerable. It is most appropriately used with peaks of small area (Figure 9-2C).

Measurement Techniques and Line Widths

In Section 1.5 the instructions for reading a buret point out that the finite width of the lines printed on a buret must be taken into account in careful work. Similarly, recorder tracings, lines on rulers, and needles on meters all have finite width, and more precise results are obtained when these line widths are correctly taken into account.

When measuring with a ruler, work consistently from line edges, which are more easily located than line centers. Bring the left edge of the zero line on the ruler up to the right edge of the line being measured. Then read the distance on the ruler corresponding to the point where the

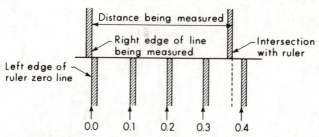

FIGURE 9–4. Placement and reading of a ruler (line thicknesses and spacing magnified for clarity).

right edge of the second line meets the ruler. In Figure 9-4 the distance is between 0.3 and 0.4 ruler units; the correct reading is 0.37.

Overlapping Peaks

In the retrieval of quantitative information from chart recordings, overlapping peaks frequently present a problem. For a peak appearing as a shoulder on a larger peak, the outline of the large peak should be sketched in at a position where it would be expected to fall if the smaller peak were absent (Figure 9-5A). The sketched-in outline then becomes a steeply changing base line for the smaller peak. In determination of the area of the smaller peak, perimeter techniques are preferable to height–width measurements because of the distortion of both height and width. With improved resolution, as in Figure 9-5C, the outline of the overlapping peak can be completed in the normal way. Intermediate cases (Figure 9-5B) are the most difficult to handle; either method A or C can be chosen.

Procedure (median time 3.2 hr)

Obtain a set of prerecorded gas chromatograms. Measure the areas of the specified peaks by planimeter and height–width techniques and finally by weighing the cut-out peaks. To avoid prejudice, make all measurements as independent of each other as possible. Take line widths into account as described above and illustrated in Figure 9-4.

Planimeter Method

Ask the laboratory instructor for instruction in setting up and operating the planimeter. Do not change the setting of the tracer arm during measurements; for greatest precision it should be the minimum that will just permit tracing the perimeter of the largest peaks.

Locate a clear space on the set of peaks provided. With a sharp pencil draw a triangle with a base of about 12 cm and a height of about 10 cm. Measure the base and height carefully to the nearest 0.1 mm. Calculate the area of this triangle by multiplying one half the height by the base.

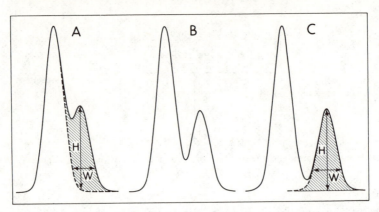

FIGURE 9–5. Pairs of peaks with resolution values of (A) 0.5, (B) 0.75, and (C) 1.0.

Test the reproducibility of the starting point of the planimeter by placing the pointer at a starting point on the triangle and recording the reading. Traverse the perimeter clockwise until the starting point is reached; then carefully reverse direction and trace the perimeter counterclockwise until the starting point is regained. The difference between this and the initial reading should be close to zero; if it is not, review your technique and repeat the procedure.

Make replicate measurements of the area of the triangle and of each peak. To measure a peak, first draw an appropriate base line. Where peaks overlap, sketch in individual peak outlines where they reasonably would be expected to lie (Figure 9-5). Tape the sheet with the peak to a smooth surface so that the planimeter wheel runs either only on or only off the paper. Place the pointer approximately in the center of the peak, with the pole arm of the planimeter about 90° to the tracer arm. Then select a point on the peak outline at which the two arms again form a right angle; mark and use this as the starting point. Record the reading on the drum and the index wheel. Trace the outline of the curve by following the center of the line, ending at the starting point. Read the planimeter again. The difference in the readings is the peak area in arbitrary units. From the planimeter reading for the triangle and its calculated area, convert each of the averages into square centimeters.

Height–Width Method

With a sharp pencil, draw a line perpendicular to the base line from the peak maximum. For an overlapping peak, extend a perpendicular line from the maximum to intersect the sketched-in outline of the associated peak (Figure 9-5). Using a ruler with a millimeter scale, measure the distance from the upper edge of the base line to the upper edge of the peak tracing at its maximum. Record this value to the nearest 0.1 mm. Next measure the peak width at one quarter of the distance from the base line to the peak maximum. The one-quarter position can be located either by direct measurement from the peak base or by use of a template (Figure

9-3). If a template is used, position it on top of the chromatographic peak, so that its top line crosses the peak maximum and its base line coincides with the peak base line. For an overlapping peak the base line of the template should be parallel to the base line of the associated peak and should pass through the peak maximum. Without moving the template, place a ruler parallel to the base line (using the grid lines on the template as a guide) so that it coincides with the intersection of the template 0.25 line with the perpendicular line, and measure the width.

Calculate the area for each peak by multiplying 0.753 by the height-width product. Pay due attention to significant figures. Report the areas in square centimeters.

Cut-and-Weigh Method

Identify each peak and cut out the specified peaks at the center of the lines with scissors.[1] Where peaks overlap, it may be possible to cut out only one of the two peaks. Handle the paper only with clean, dry hands. Next, cut out the standard triangle. Weigh the peaks and the triangle on an analytical balance, and record the weights in grams.

Convert the peak weights to square centimeters. Again, in making calculations use the correct number of significant figures; Figure 9-2 is helpful in this regard.

PROBLEMS

9–3.   The relative standard deviation in the peak area measured by the height–width technique for a 30-cm^2 peak having a height-to-width ratio of 10 is 0.6%. What error would be expected for a peak with an area of $1.2 \ cm^2$ and the same height-to-width ratio? If the planimeter technique were used, the 30-cm^2 peak would be measured more precisely (0.5%). What error would be expected for the 1.2-cm^2 peak by this method?

9–4.   What would be the effect on peak shape and the precision of measurement for (a) sharp peaks and (b) flat peaks if the recorder chart speed were decreased? If the recorder sensitivity were increased?

SELECTED REFERENCES

D. L. Ball, W. E. Harris, and H. W. Habgood, *Anal. Chem.* **40**, 129 (1968); **40**, 1113 (1968); *J. Gas Chromatog.* **5**, 613 (1967).

J. Gas Chromatog. **5**, 595 (1967). Results of a survey on methods of evaluating data obtained by gas chromatography.

---

[1] Cut around the peak in the direction that minimizes handling of the peak. For right-handed persons, this is clockwise, contrary to natural inclination.

Do not cut out peaks until satisfactory results have been obtained by the height–width and planimeter methods.

Chapter 10

SEPARATIONS IN ANALYSIS: SINGLE-STAGE AND BATCH TECHNIQUES

10-1. GENERAL BACKGROUND

The experiments to this point emphasize analytical measurements without extensive prior operations other than dissolution. Interferences were avoided either by selective methods of analysis or the addition of reagents that blocked interference through complexation, changes in oxidation state, and so on. In many cases, however, interferences from other substances present in a sample cannot be prevented, and so a separation of these interfering substances from the material of interest becomes necessary. The relation between the separation step and the other major steps of the overall analytical operation is shown in Figure 10-1.

In this chapter, one approach to the problem of removing interferences prior to the measurement step, that of a single-stage operation, is illustrated in several ways. Single-stage separations are recommended whenever the separation factor is sufficiently favorable that the material of interest can be isolated from the interference in one or two operations. The *separation* (or *enrichment*) *factor* is defined as the relative change, as a result of the separation, in the ratio of the quantity of an unwanted substance to that of a desired component. A separation involves the formation of two phases, with the wanted material concentrated in one and the unwanted in the other. Common separation techniques include precipitation, extraction from one liquid phase into a second immiscible one, and distillation. If the technique involves physical isolation of the two phases from each other before a second equilibration is carried out, it is called a batch technique. If the separation factor is small, one or perhaps two operations may be sufficient to provide the separation

198

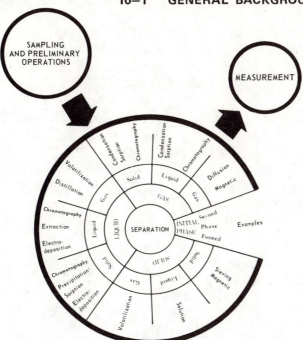

FIGURE 10–1. The location of the separation step in the analytical operation, with examples of various techniques.

required. If it is near unity, the single-stage technique may have to be repeated on a sample a number of times, or continuous multiple operations may be preferable (Chapter 11).

The experiments in this chapter include the liquid–liquid extraction of two metal complexes [nickel dimethylglyoxime into chloroform and iron(III) chloride into methyl isobutyl ketone] and two special types of precipitation separations (a precipitation of aluminum from homogeneous solution and a carrier precipitation of thorium at the picogram level). The experiments also encompass a variety of measurement procedures designed to broaden the range of measurement techniques.

Liquid-Liquid Extraction of Metal Complexes

Liquid-liquid extraction procedures are used widely for the separation of compounds and for the concentration of small amounts of substances. Solvents immiscible in water are generally less polar than water and so tend to dissolve nonpolar solutes to a greater extent. If an ion, say a metal ion, is converted to an uncharged species, its solubility in most organic solvents is enhanced and its extraction from water may then be possible. Uncharged species can be formed either directly by complexation with negatively charged ligands,

$$M^{n+} + nL^- \rightleftarrows ML_n \qquad (10-1)$$

or by displacement of protons from coordination sites on ligands,

$$M^{n+} + nHL \rightleftarrows ML_n + nH^+ \qquad (10-2)$$

An example of Equation (10-2) is the complex formed between magnesium and 8-hydroxyquinoline: (10-3)

Uncharged species can form also when a large ion of low charge is extracted along with an ion of opposite charge into an organic phase as an ion pair. As an example, cobalt(III) thiocyanate can be extracted quantitatively into chloroform as an ion pair with the tetraphenyl-arsonium cation, $(C_6H_5)_4As^+$:

$$Co^{3+} + 4SCN^- \rightleftarrows Co(SCN)_4^- \qquad (10-4)$$

and

$$(C_6H_5)_4As^+Cl^- + Co(SCN)_4^- \rightleftarrows [(C_6H_5)_4As^+Co(SCN)_4^-] + Cl^- \qquad (10-5)$$

Still another way in which uncharged species are produced is through formation of an acid whose conjugate base is a large anion of low charge. These acids often can be readily extracted into dipolar organic solvents. One example of a system of this kind is the extraction of iron (III) as $HFeCl_4$ from an aqueous hydrochloric acid solution into ethyl ether, $C_2H_5OC_2H_5$:

$$4Cl^- + Fe^{3+} \rightleftarrows FeCl_4^- \qquad (10-6)$$

and

$$FeCl_4^- + H^+ \rightleftarrows HFeCl_4 \qquad (10-7)$$

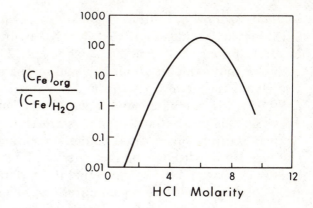

FIGURE 10–2. Distribution of iron(III) between ethyl ether and water as a function of hydrochloric acid concentration in the water.

The extractibility of a substance can be expressed in terms of the *distribution ratio,* defined as the ratio at equilibrium of the concentration of the substance in the nonaqueous phase to its concentration in the aqueous phase. The distribution of iron between water and ether goes through a maximum at a hydrochloric acid concentration of about 6 *M* (Figure 10-2). The decrease at higher hydrochloric acid concentrations has been attributed to increased solubility of ether in the aqueous acid, which appreciably increases the volume of the aqueous phase relative to the nonaqueous phase. Since the distribution ratio is only about 100 under optimum conditions, several extractions are necessary to remove the iron

quantitatively. Methyl isobutyl ketone (MIBK), $CH_3\overset{\overset{O}{\|}}{C}CH_2CH(CH_3)_2$, extracts iron as effectively as does ethyl ether, but has a more favorable distribution ratio and is less flammable.

Precipitation from Homogeneous Solution

Besides liquid–liquid extraction, numerous other means are available for separating a substance of interest from interferences. One such method is selective, or fractional, precipitation.

Gravimetric analysis requires low solubility of the precipitated compound. Low precipitate solubility, however, makes it more difficult to keep the supersaturation ratio low, and consequently coprecipitation of contaminants may become serious. One way to reduce contamination of a precipitate is to generate a precipitating agent slowly and uniformly throughout the solution containing the sample. Experiment 10-4 illustrates an application of this technique to the determination of aluminum; hydroxide is generated to precipitate aluminum hydroxide from a mixture of aluminum, zinc, and magnesium salts. This system provides a convenient illustration of precipitation from homogeneous solution.

Iron(III), aluminum, zinc, and magnesium form insoluble hydroxides, all of which precipitate upon direct addition of base. Their solubilities, however, differ. Calculations based on solubility products of individual hydroxides show that iron(III) hydroxide begins to form from 0.1 M solutions of the metals at about pH 2.3, aluminum hydroxide at 3.4, zinc hydroxide at about 5.5, and magnesium hydroxide at 9.0. For the aluminum in a 0.1 M solution to be precipitated quantitatively (>99.9%), the aluminum ion concentration must be reduced to $10^{-4} M$. To attain this concentration requires a pH of 4.8; thus the aluminum can be quantitatively precipitated before zinc or magnesium hydroxide begins to form. Iron hydroxide, on the other hand, precipitates before aluminum hydroxide, but at a pH so close that it interferes in a direct precipitation. Iron, therefore, must be separated prior to aluminum precipitation. This separation is readily accomplished by extraction of the iron into an immiscible solvent as described earlier.

The close control of pH required in selective hydroxide precipitation is made possible by use of a succinate-succinic acid buffer. Succinic acid is a dibasic acid ($pK_1 = 4.15$, $pK_2 = 5.60$). A solution of this acid when half neutralized has a pH of about 4.9; it will maintain this pH to within about a tenth of a unit unless so much acid or base is added that the capacity of the buffer is exceeded.

If a solution containing succinic acid and aluminum is half neutralized by direct addition of base, the pH will rise rapidly through the region 3.4 to 4.8. The supersaturation ratio for aluminum hydroxide will be enormous, and it will form as a gelatinous precipitate. Coprecipitation of other metal ions present with the aluminum will be extensive, and the aluminum can be obtained in pure form only by filtering, washing, redissolving, and reprecipitating the aluminum hydroxide. Coprecipitation can be minimized, however, and a pure precipitate obtained in a single step by slow, homogeneous generation of hydroxide ions in the solution. This general process of precipitant generation is called precipitation from homogeneous solution.

An effective way to raise the pH of a solution slowly is to add urea to the solution and then heat it. Urea solutions slowly hydrolyze on warming to form ammonia and carbon dioxide:

$$(NH_2)_2CO + H_2O \rightarrow CO_2 + 2NH_3 \qquad (10-8)$$

The carbon dioxide, not being soluble in hot acid solutions, escapes. The ammonia remains and raises the pH by neutralizing part of the succinic acid. Because the pH increases slowly over several hours, the supersaturation ratio never becomes high, and the nucleation rate remains low. These conditions are favorable for the growth of particles already in existence so that a relatively dense, pure precipitate is obtained. Limiting the amount of urea added prevents the pH from rising to levels where the solubility products of zinc and magnesium hydroxide are exceeded.

Carrier Precipitation

Coprecipitation is defined as the contamination of a precipitate by substances that normally are soluble (Section 5-1). It usually is regarded as a source of error and therefore undesirable. An exception is in ultratrace analysis, where the minute amounts of material make conventional techniques infeasible and require special procedures to be devised. In radiochemistry, for instance, the measurement technique, radioactivity counting, is so sensitive that a single atom of an element can be detected. Consider as an example the problem of separating and analyzing the small amount of thorium-234 in radiochemical equilibrium with a solution of a uranium salt.

When base is added to a solution of a thorium[1] salt at ordinary concentrations, thorium hydroxide precipitates readily. At exceedingly low thorium concentrations, however, nucleation cannot occur and thorium hydroxide cannot be recovered by filtration.

One technique of separating thorium-234 from uranium is that of making coprecipitation work in a positive way. Gelatinous precipitates are highly susceptible to coprecipitation of foreign ions because the large surface areas promote extensive adsorption. Iron(III) hydroxide ordinarily precipitates in a gelatinous form. If a trace of an iron(III) salt is added to a thorium solution and iron(III) hydroxide is precipitated rapidly by the addition of base, an extremely high supersaturation ratio is produced, resulting in an iron(III) hydroxide precipitate with a large surface area. The tendency for thorium hydroxide to adsorb on this surface is great; more than 95% of the thorium ends up on the iron(III) hydroxide precipitate and remains there even after filtration and washing. A precipitate such as iron(III) hydroxide used in this way is called a *scavenger* precipitate.

10–2. LIQUID-LIQUID EXTRACTION AND SPECTROPHOTO-METRIC ANALYSIS: DETERMINATION OF TRACE NICKEL IN COPPER

Background

In this experiment a small amount of nickel is separated from a large quantity of copper by extraction of nickel dimethylglyoxime, $Ni(DMG)_2$, from water into chloroform. Although a single extraction does not quantitatively separate all the nickel, high-quality results can be obtained

---

[1] The only geologically stable thorium isotope is thorium-232, with a half-life of 1.39 × 10^{10} years.

by carrying standards through the same procedure and preparing a calibration curve. Beer's law is followed by $Ni(DMG)_2$ in chloroform over a wide range of concentration, so a graph of absorbance against concentration should be linear. The absorbances of all solutions should be read relative to a blank prepared by carrying a portion of deionized water through the extraction and sample treatment. In this way nickel impurities in the reagents are compensated for.

Thiosulfate is added prior to extraction to form an anionic complex, $Cu(S_2O_3)_2^=$, that is not extracted into chloroform. Similarly, tartrate is added to complex any iron present. Hydroxylamine hydrochloride is added to prevent oxidation of $Ni(DMG)_2$ to a nickel(IV) complex with dimethylglyoxime, which has a different absorption spectrum.

Procedure

Preparation of Solutions

Accurately weigh into a 100-ml volumetric flask enough nickel powder to prepare 100 ml of approximately 0.04 M solution. Add 15 ml of 6 M HNO_3 and heat on a hot plate until dissolution is complete. Do not stopper the flask! Neutralize with 6 M NaOH to the first appearance of a permanent precipitate of nickel hydroxide. Add a few drops of 6 M acetic acid to clear the solution. Dilute to volume, mix, and then pipet a 10-ml aliquot into a 250-ml volumetric flask and dilute to volume to give a standard solution containing about 100 ppm of nickel.

Prepare a buffer of pH 6.5 by diluting 8.7 ml of 6 M acetic acid to 100 ml and then adding 10 ml of this solution to 40 ml of water containing 15 g of sodium acetate.

Obtain a nickel sample, dissolve if necessary, and dilute to volume in a 100-ml volumetric flask. The sample solution should contain in the range of 1 to 4 mg of nickel per 100 ml.

Analytical Procedure

Clean six 25- by 150-mm test tubes and deliver 1, 2, and 3.5 ml of standard nickel solution from a buret into three of them, recording the exact volume in each case. Dilute each standard to approximately 10 ml. Pipet 10 ml of the unknown sample solution into each of two more test tubes. Add 10 ml of water to the remaining tube to serve as a blank. To each test tube add in succession (with vigorous swirling after each addition) 0.4 g of sodium hydrogen tartrate,[1] 5 ml of pH 6.5 buffer, 2.5 g

---

[1] Sodium hydrogen tartrate is slow to dissolve. Dissolution of material remaining after all the reagents have been added can be hastened by gentle warming.

of sodium thiosulfate, 1 ml of 10% hydroxylamine hydrochloride in water, and 2 ml of 1% dimethylglyoxime in ethanol.

Pipet 10 ml of chloroform into each test tube and mix the two phases thoroughly by drawing portions of the aqueous layer into a large medicine dropper and squirting them into the chloroform. A 10-ml Mohr pipet works well for this purpose. Stopper the tubes to minimize evaporation of chloroform. Allow the phases to separate, then with the pipet transfer most of the chloroform layer to a 10-cm qualitative filter paper of medium porosity, and collect the filtrate in a spectrophotometer sample tube; again stopper the tubes after filtration. (Follow the procedure outlined in Experiment 6-3 for the selection and use of matched tubes for the spectrophotometer.)

Determine the absorption spectrum of the $Ni(DMG)_2$ complex in chloroform by measuring the intermediate standard nickel solution. (This procedure also is outlined in Experiment 6-3.) Record readings at 20-nm intervals from 350 to 650 nm, using the extracted blank solution to set the instrument to zero absorbance at each wavelength. To define the spectrum more clearly, take readings every 5 nm in the region from 400 to 450 nm. Plot absorbance (vertical axis) against wavelength (horizontal axis) and select the wavelength to be used for analysis; confirm this value with your laboratory instructor. Then measure the absorbance of the other standards and the two sample solutions. Repeat the measurements on each solution several times.

Calculations

Plot the absorbance readings of the standard nickel solutions (vertical axis) against the nickel concentration in milligrams per milliliter (horizontal axis). Draw the best straight line through the points; then read from the line the concentrations corresponding to the absorbance values of the two sample solutions. Calculate and report the total milligrams of nickel in the original sample.

PROBLEMS

10–1.   A 0.1548-g sample of an alloy was dissolved and analyzed for nickel by the procedure of this experiment. The weight of nickel taken for the standard was 0.2407 g, and 1.02, 2.07, and 3.46 ml of standard solution gave absorbance readings of 0.140, 0.288, and 0.476. If the absorbance of the sample was 0.350, what was the percentage of nickel in the alloy?

10–2.   What are the advantages of a calibration curve over comparison with a single standard as in Experiment 6-3? Which method would be the choice for samples covering a range of concentrations? Which would be best for quality control of a single product of unvarying composition?

SELECTED REFERENCES

A. K. De, S. M. Khopkar, and R. A. Chalmers, *Solvent Extraction of Metals,* Van Nostrand Reinhold, London, 1970. A survey of liquid-liquid extraction methods for application to the determination of metals. Contains many references to the original literature.

G. H. Morrison and H. Frieser, *Solvent Extraction in Analytical Chemistry,* Wiley, New York, 1957.

W. Nielsch, *Z. Anal. Chem.* **150,** 114 (1956). A study of the determination of small amounts of nickel in copper salts by the method of this experiment. (In German)

10–3. LIQUID–LIQUID EXTRACTION: SEPARATION OF IRON FROM ALUMINUM. DETERMINATION OF ALUMINUM BY EDTA

Background

For background on liquid–liquid extraction, refer to Section 10-1. In the method given here the analysis is completed by EDTA titration. A gravimetric finish by precipitation from homogeneous solution is discussed in Experiment 10-4. The extraction part of the two procedures, however, is the same.

EDTA Determination of Aluminum

A general discussion of EDTA–metal complexes is given in Experiment 3-3. A source of difficulty in the determination of aluminum with EDTA is that formation of the aluminum–EDTA complex takes place only slowly, especially if the aluminum is present in the form of hydroxide complexes. Therefore conditions must be carefully controlled if titration of this element with EDTA is to be successful.

The formation constant for the reaction $Al^{3+} + Y^{4-} \rightleftarrows AlY^-$ ($K = 10^{16.1}$) does not indicate the extent of formation of AlY^- in water, because hydrogen ions from water compete with aluminum for sites on the EDTA anion and hydroxyl ions from water compete with EDTA for sites in the coordination sphere of aluminum. The pH of the solution, then, determines to a large degree the fraction of the total aluminum in a sample that is present in the EDTA complex.

To facilitate calculation of the extent of complex formation at various pH values, chemists have introduced the idea of *conditional formation constants*. These are formation constants that apply under specified conditions, usually of pH. For example, the formation constant for zinc with EDTA is $10^{16.5}$; it may be written

$$\text{Zn}^{++} + \text{Y}^{4-} \rightleftarrows \text{ZnY}^{=} \qquad K_f = 10^{16.5} = \frac{[\text{ZnY}^{=}]}{[\text{Zn}^{++}][\text{Y}^{4-}]}$$

Inspection of the acid dissociation constants for EDTA, however, shows that Y^{4-} exists as a significant fraction of the total amount of EDTA present only when the pH is 10 or greater. Below pH 10, protonated forms of EDTA predominate. Also, zinc forms a hydroxide complex, Zn(OH)^{+}, of appreciable stability ($K_f = 10^{4.3}$) that should be taken into account; if the pH becomes too high, a precipitate of Zn(OH)_2 ($K_{sp} = 5 \times 10^{-18}$) forms. The conditional formation constant gives the relation between EDTA and zinc as

$$K_{cond} = \frac{[\text{ZnY}^{=}]}{C_{\text{Zn}} C_{\text{EDTA}}}$$

where C_{Zn} is the molar concentration of zinc present in all forms other than $\text{ZnY}^{=}$, and where C_{EDTA} is the molar concentration of EDTA present in all forms other than $\text{ZnY}^{=}$. This equation is useful in determining whether a reaction has a sufficiently large equilibrium constant under the *conditions used* to be quantitative, that is, whether the ratio of C_{Zn} to $\text{ZnY}^{=}$ is 0.001 or less at the equivalence point.

Conditional formation constants are calculated from the constants for all the equilibria involved in the system and often are shown graphically as functions of pH. Plots of pH against the log of the conditional formation constants for the zinc-EDTA and aluminum-EDTA systems are shown in Figure 10-3. Note that for zinc the conditional constant reaches a maximum at about pH 9; at lower pH values hydrogen ions from water compete successfully with zinc for sites on the EDTA anion, whereas at higher pH values complex formation between zinc(II) and hydroxide ion becomes predominant. With aluminum, complexation by hydroxide is much stronger (for example, K_f for $\text{AlOH}^{++} = 10^{8.8}$), and so the conditional constant for aluminum with EDTA begins to decrease at lower pH values.

The minimum conditional formation constant necessary for quantitative reaction at the part-per-thousand level is about 10^6 Figure 10-3 indicates that, if we wish to utilize aluminum–EDTA complex formation for a quantitative determination of aluminum, we must operate within a

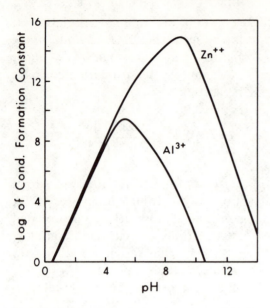

FIGURE 10–3. Plots of the logarithm of the conditional formation constants for aluminum(III) and zinc(II) with EDTA as a function of pH.

rather narrow pH range around pH 5. Zinc, on the other hand, can be titrated over the much wider range of pH 4 to 12.

Differences in conditional formation constants sometimes can be put to direct use. For instance, zinc can be titrated at pH 10 or 11 without interference from aluminum because at these high pH values the conditional constant for zinc is still large, whereas that for aluminum is sufficiently small that essentially none of the aluminum–EDTA complex is formed.

In this experiment a solution containing aluminum, iron, zinc, and magnesium is analyzed for aluminum. First, the iron is removed by solvent extraction from aqueous hydrochloric acid into methyl isobutyl ketone. Next, the remaining mixture of aluminum, zinc, and magnesium is treated with EDTA in excess of the amount required to complex the aluminum and zinc present. The pH is adjusted to 5 with hexamethylenetetramine, at which value the magnesium complex of EDTA does not form appreciably. The excess EDTA then is titrated with zinc nitrate solution. At this point all the EDTA is present as a complex with either zinc or aluminum. An excess of fluoride ion is added and the solution boiled a minute or two. Fluoride forms a stable complex with aluminum (overall equilibrium constant for $AlF_6^{3-} = 10^{20}$) but not with zinc, and thus it will displace EDTA from the coordination sphere of aluminum. The amount of EDTA released is equivalent to the amount of aluminum in the sample. This EDTA is titrated with standard zinc solution; the number of moles of zinc required in this titration is equivalent to the moles of aluminum in the sample.

The use of fluoride to displace aluminum in this way is an example of a type of *masking* reaction. A masking agent in complexation analysis is an auxiliary complexing agent added to a mixture of metal ions to selectively bind one or more of the metals present and prevent their reacting with the titrant. A more usual application of masking in the zinc–aluminum system is the determination of the amount of zinc in a mixture of the two by addition of fluoride to complex the aluminum so that zinc, which does not form a stable fluoride complex, can be titrated directly. Another example is the addition of cyanide to a mixture of bismuth and iron. Iron(III) forms a cyanide complex of such stability that bismuth can be titrated readily with EDTA in acid solution without interference.

Masking agents are used also in analytical methods other than complexation titrations. For example, in the determination of nickel in mixtures of iron and nickel through precipitation of nickel dimethyl-glyoxime, tartrate is added to prevent precipitation of iron hydroxide under the slightly alkaline conditions employed.

Hexamethylenetetramine[1] is chosen to adjust the pH, for two reasons. First, it is a sufficiently weak base that the presence of a local excess during its addition will not cause hydrolysis of aluminum. Hydrolysis occurring when a stronger base such as ammonia is used results in formation of polynuclear complexes containing Al-O-Al bonds that are

---

[1] Hexamethylenetetramine is an unusual polycyclic cage molecule containing four nitrogen atoms, each connected to the other three by a $-CH_2-$(methylene) bridge. The compound is inexpensive, being readily formed by heating a mixture of formaldehyde and ammonia. It has been used for a variety of purposes, ranging from an intermediate in the production of Bakelite plastics and RDX high explosives to a urinary antiseptic.

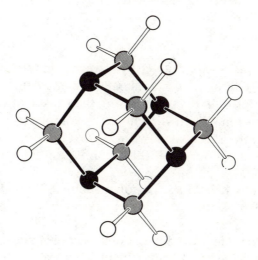

slow to revert back to mononuclear forms even in strongly acidic solution. The presence of these species blocks complexation of aluminum by EDTA. Consequently, a strong base should not be used to adjust the pH of an acidic solution containing aluminum prior to EDTA titration. Second, the pK_b of hexamethylenetetramine, 8.9 (pK_a of $HB^+ = 5.1$) provides effective buffering at pH 5, the optimum pH value for formation of the aluminum–EDTA complex.

The indicator used in this titration, Xylenol Orange, is an acid–base as well as a complexation indicator. The structure is

In solutions of pH below 5.4, the compound is yellow, and above 7.4 it is red. Since the zinc complex also is red, a change in indicator color in zinc titrations can be seen only in solutions of pH 5.4 or below. Aqueous solutions of Xylenol Orange are stable for only a week or so; use of older solutions results in poor, drawn-out end points.

Procedure (median time 5.1 hr)

Preliminary Operations

Preparation of 0.025 M Zn(NO$_3$)$_2$. Weigh to the nearest 0.1 mg approximately 0.82 g of pure granulated zinc metal into a 100-ml beaker. Add slowly 5 ml of 6 M HNO$_3$, covering the beaker with a watch glass to avoid loss of solution. When dissolution is complete, rinse the liquid on the watch glass into the beaker with 10 to 15 ml of water, transfer the contents of the beaker quantitatively to a 500-ml volumetric flask, fill to the mark, and mix.

Preparation of 0.05 M EDTA. Dissolve about 9.3 g of the disodium salt of EDTA, Na$_2$H$_2$Y·2H$_2$O, in 500 ml of water. This solution does not need to be standardized.

Preparation of Sample. Obtain a liquid sample, dilute to volume in a 100-ml volumetric flask, and mix. Pipet 10-ml aliquots of this solution into each of three 25-mm test tubes and add 12 ml of 12 M HCl to each.

Extraction of Iron(III)

Add a few milliliters of 6 *M* HCl to about 100 ml of methyl isobutyl ketone (MIBK) in a 200-ml conical flask and shake[2] to saturate the MIBK with the aqueous phase. Add about 10 ml of the HCl-saturated MIBK to one sample solution.[3] Draw a portion of the MIBK layer into a large medicine dropper or 10-ml Mohr pipet and squirt it into the aqueous layer.[4] Repeat this operation 10 to 15 times and then allow the phases to separate.[5] Withdraw the MIBK layer with the pipet and transfer it to an 18-mm test tube. Repeat the extraction with fresh portions of MIBK until all the yellow iron(III) chloride has been removed from the water layer.[6] Combine all the MIBK extracts from a single aliquot in one tube. After the last extraction, wash traces of aqueous solution from the upper surfaces of the tube with a small portion of MIBK, and add this portion to the collected MIBK extracts.

Scrub aluminum from the MIBK extracts by adding about 2 ml of 6 *M* HCl and mixing as above. Carefully remove and discard the MIBK layer. Take care that none of the aqueous layer is removed and discarded along with the MIBK, or aluminum will be lost. Wash the HCl fraction with two more portions of MIBK, then combine the aqueous layers from the extraction and scrubbing operations in a 100-ml volumetric flask, dilute to volume, and mix.

Repeat the extraction process on the second 10-ml portion of original sample and place in a second 100-ml volumetric flask. Finally, repeat on the third portion to provide a third extracted aliquot.

Titration of Aluminum

Pipet 10-ml portions of the extracted solution from one of the volumetric flasks into 200-ml conical flasks. Add approximately 25 ml of

---

[2] Do not waste this solution. Here, as in all experimental work, take only the amounts specified from the reagent stock. Not only are many chemicals expensive, but problems of disposal and pollution are aggravated when excessive amounts are taken, only to be dumped down the drain.

[3] At this point, one sample should be carried entirely through the extraction operations to provide practice in the technique. High concentrations of MIBK vapors are irritating to the eyes and mucous membranes, so use this solvent with adequate ventilation.

[4] This operation is most conveniently carried out by placing the tip of the pipet in the upper phase, a half centimeter or so above the interface of the two liquids, and squeezing the pipet bulb while holding the pipet stationary. The two phases will be quickly and efficiently mixed. The pipet bulb must fit firmly on the end of the pipet, or leakage of air during the squeezing process may force liquid up the pipet into the bulb and cause loss of aluminum.

[5] After equilibrium has been established in the first extraction, the color of the initially yellow water layer will decrease in intensity, and that of the initially clear MIBK layer will become yellow or slightly orange. The mixing should be continued until no further change is seen in the colors of the two phases. After the second extraction the aqueous layer may turn a faint green or brown due to impurities. This will not affect the extraction and may be ignored.

[6] The third MIBK layer should be colorless or nearly so. If not, do another extraction.

0.05 M EDTA solution to each flask and swirl to mix; then add 10 ml of 3 M hexamethylenetetramine solution, and boil for 1 to 2 min to ensure complete formation of the aluminum-EDTA complex. Remove from heat, add 4 to 5 drops of 0.2% Xylenol Orange indicator solution, and titrate the hot solution with 0.025 M Zn^{++} to a definite color change. Add 25 ml of a saturated solution of sodium fluoride, swirl to mix, and boil again vigorously for 1 to 2 min. Remove from heat, add 2 to 3 more drops of Xylenol Orange indicator, and immediately titrate again.[7]

Repeat this procedure on sets of aliquots taken from the volumetric flasks containing the second and third extraction. Calculate and report the weight in grams of aluminum in the entire sample.

Calculations

Calculate the weight of aluminum present in each titrated aliquot from the molarity of the standard zinc nitrate solution and the volume of zinc solution required for titration from the first to the second Xylenol Orange end point for each aliquot. Then calculate the total weight in grams of aluminum in the original sample.

PROBLEMS

10–3.   A sample of aluminum analyzed by the procedure of this experiment required 9.73 ml of zinc nitrate solution to titrate the EDTA released from aluminum through the addition of fluoride. If the molarity of the zinc solution was 0.0500 M, what was the total weight of aluminum present in the initial sample?

10–4.   Write the chemical equations for each of the steps in the determination of aluminum by the procedure of this experiment.

SELECTED REFERENCES

G. H. Morrison and H. Freiser, *Solvent Extraction in Analytical Chemistry,* Wiley, New York, 1957. A valuable introduction to the use of liquid-liquid extraction in separations.

D. D. Perrin, *Masking and Demasking of Chemical Reactions,* Wiley, New York, 1970. Discusses principles of masking reactions and provides many examples of applications.

R. Pribil and V. Vesely, *Talanta* 9, 23 (1962). A study of the determination of aluminum by complexation. Includes a critical investigation of the use of fluoride to determine aluminum by displacement of a chelating agent.

---

[7] The volume of zinc nitrate solution required for calculation of the aluminum present is that required to go from the end point of the first titration to that of the second.

10–4. LIQUID-LIQUID EXTRACTION: SEPARATION OF IRON FROM ALUMINUM. GRAVIMETRIC DETERMINATION OF ALUMINUM BY PRECIPITATION FROM HOMOGENEOUS SOLUTION

In this experiment the analysis of aluminum in the same mixture of salts as in Experiment 10-3 is performed, except that precipitation from homogeneous solution is used to separate aluminum from zinc and magnesium. The analysis step involves filtration of the precipitated aluminum hydroxide, followed by ignition and weighing as Al_2O_3. The iron is separated from the mixture by liquid–liquid extraction as in Experiment 10-3. The principles of liquid–liquid extraction of metal complexes and of precipitation from homogeneous solution are discussed in Section 10-1, and the experimental techniques of gravimetric analysis in Section 5-1.

Procedure (median time 7 hr)

Preliminary Operations

Obtain a sample, dilute to volume in a 100-ml volumetric flask, and mix. Pipet 10-ml aliquots of this solution into each of three 25-mm test tubes and add 12 ml of 12 M HCl to each.

Extraction of Iron(III)

Extract the iron(III) from each of the aliquots by the procedure given under **Extraction of Iron(III)** in Experiment 10-3, with the minor modification given below. When the extractions are complete, combine the aqueous layers from the extraction and scrubbing operations for each aliquot in each of three 250-ml beakers (rather than in 100-ml volumetric flasks as specified in Experiment 10-3).

Precipitation of Aluminum

Add to the solution in each 250-ml beaker a few grains of $NaHSO_3$ [1] and a few drops of thymol blue indicator. Cover with a watch glass, and add 6 M NH_3 dropwise with constant stirring until the indicator turns

---

[1] Sodium hydrogen sulfite, $NaHSO_3$, is added to reduce oxidizing impurities in the system that otherwise would destroy the thymol blue indicator.

yellow, then 5 to 10 drops more.[2] Add 3 g of succinic acid, 5 g of NH_4Cl, and 3 g of urea.[3] Stir until most of the solids dissolve.[4] Remove the stirring rod, cover tightly with a watch glass, and leave overnight on a hot plate set at low heat. Filter through medium-porosity paper (Section 5-1 and Figure 5-2). Test for completeness of precipitation by adding a few drops of NH_3 to the clear filtrate. Prepare about 500 ml of a 1% succinic acid wash solution made neutral to methyl red with NH_3, and wash the precipitate five to ten times. Finally, wash the filter paper with distilled water until a test for chloride in the filtrate is negative.

A small amount of precipitate will adhere to the beaker walls. Loosen as much of it as possible with a policeman and transfer to the filter paper; add about 1 ml of dilute HCl to the beaker to dissolve the remainder.[5] Add 10 ml of water, then a drop of methyl red, and neutralize with NH_3. Transfer this precipitate to the filter paper with the 1% succinate wash solution.

Place the aluminum precipitate and filter paper in a crucible ignited previously to constant weight (± 0.2 mg). Position the crucible at an angle of about 45° (Figure 5-3) and about 2 cm above a cool flame about 1 cm high. Heat gently at first, then more strongly after the paper has charred. Finally, heat at full burner flame for 20 to 30 min. Cool 30 min, weigh, and reheat to constant weight.[6]

Report the weight in grams of aluminum in the entire sample.

PROBLEMS

10-5.   A 10.014-ml aliquot taken from a 100-ml sample containing an aluminum salt yielded 0.3412 g of aluminum oxide. What was the total weight of aluminum in the sample?

---

[2] The pH at the point of indicator change is about 2.5 to 3.0; the additional 5 to 10 drops of NH_3 raises it to 6 or 7. If insufficient NH_3 is added here, the additional amount generated by urea hydrolysis will not raise the pH enough to precipitate all the aluminum. If a precipitate appears, it usually will dissolve when succinic acid is added in the next step. If dissolution is not complete after the addition of succinic acid and mixing, add 6 M HCl dropwise with stirring until no solid remains, and then continue.

[3] The amount of urea used should be sufficient to raise the pH upon hydrolysis to about 5. If no precipitate appears after heating, reneutralize with NH_3, add another 5 g of urea, and heat again for 24 hr.

[4] Succinic acid dissolves slowly.

[5] Although most of the precipitate will be granular and coarse, a small amount will form a thin, adherent film on the beaker walls that is difficult to remove with a policeman. This can be dissolved with a little HCl and reprecipitated with ammonia. Since the amount of precipitate is small and interfering ions are no longer present, direct reprecipitation is acceptable.

[6] Use a small flame during the early stages of ignition; do not rush the process. If the precipitate and filter paper are ignited too rapidly at a high temperature, some elemental carbon may form that is hard to remove. Also, portions of charred paper, with particles of precipitate adhering to them, may be carried out of the crucible. The final precipitate should be white. Do not touch red-hot crucibles with cold crucible tongs; breakage might result.

10–6.   A silicate sample weighing 1.0671 g gave 0.2917 g of iron(III) and aluminum oxides. The mixture of oxides was dissolved, and the iron(III) reduced to iron(II) and titrated with 11.45 ml of 0.02146 M potassium permanganate. What was the percentage of Al_2O_3 in the silicate?

10–7.   For a distribution ratio of 100, how many extractions are needed to remove at least 99.9% of the iron(III) from a 10-ml aqueous aliquot into ether by the procedure of this experiment? Assume that 10 ml of ether is used for each extraction. How many extractions would be necessary if 3 ml of ether were used per extraction?

10–8.   Calculate the pH at which copper(II) hydroxide would just begin to form (a) from a 0.1 M solution of copper(II) nitrate and (b) from a 0.001 M solution.

SELECTED REFERENCES

L. Gordon, M. L. Salutsky, and H. H. Willard, *Precipitation from Homogeneous Solution,* Wiley, New York, 1959.
W. F. Hillebrand, G. E. F. Lundell, H. A. Bright, and J. I. Hoffman, *Applied Inorganic Analysis,* 2nd ed., Wiley, New York, 1953. Contains detailed procedures for gravimetric analyses of naturally occurring materials.
H. H. Willard and N. K. Tang, *J. Amer. Chem. Soc.* 59, 1190 (1937); Anal. Chem. 9, 357 (1937). Early papers on the technique of precipitation from homogeneous solution.

10–5. CARRIER PRECIPITATION: SEPARATION OF THORIUM-234 FROM URANIUM-238. THORIUM-234 BY RADIOACTIVITY MEASUREMENT.

> Nature, it seems, is the popular name for milliards and milliards and milliards of particles playing their infinite game of billiards and billiards and billiards.
>
> *Piet Hein*

Background

Most of the experiments up to this point involve chemical reactions with favorable equilibrium constants. But conditions for quantitative reaction often are not feasible, or conventional methods of calculating chemical equilibria cannot be applied. In such systems a standard and sample are handled under as nearly identical conditions as possible, and all operations are carried out in as similar a way as possible. Results are calculated by comparison of the sample with the standard. When reactions are nonquantitative, the extent of reaction is much more sensitive to slight variations in procedure than when reactions are driven to completion. The determination of thorium-234 is an example of nonquantitative recovery of a substance to be measured.

Because adsorption of thorium-234 by iron(III) hydroxide precipitate is not quantitative, experimental conditions must be carefully controlled and reproduced. A standard solution of thorium-234 is carried through the same process as the sample to compensate for incompleteness of scavenging, radioactive decay, and counting geometry. The simplest way to prepare a standard solution of thorium-234 is to determine by conventional chemical means the concentration of uranium in a solution of a uranium salt in secular equilibrium with thorium-234. The concentration of thorium in this solution then can be calculated from the concentration of uranium and the half-lives of thorium-234 and uranium-238. Since this standard solution of thorium-234 contains uranium, for reproducible separation conditions the sample also must contain uranium. Uranium-238 in the sample continuously produces thorium-234, which must be taken into account. Accordingly, the time of separation is recorded, and allowance for thorium-234 produced is made through a calculation involving the thorium-234 half-life. Since the half-life of thorium-234 is 24 days, the time of separation need not be noted to closer than the nearest hour or two. In the separation, uranium first precipitates as yellow ammonium diuranate, $(NH_4)_2U_2O_7$, upon addition of ammonium carbonate. With an excess of carbonate a soluble complex, probably $UO_2(CO_3)_3{}^{4-}$, forms and the precipitate redissolves.

Radioactivity and Thorium-234

This experiment illustrates some of the characteristics of analysis by radioactivity. Radiochemical analysis is a subject that requires knowledge of both chemical properties and nuclear behavior. For the first 103 elements there are 272 stable isotopes, 58 naturally occurring radioactive isotopes, and about 1200 artificially produced radioactive isotopes. For tracer work, radioactive isotopes of about 60 of the elements can be purchased.

The time at which a radioactive atom will decay is completely unpredictable. On the other hand, a sufficiently large number of atoms of a radioactive isotope will behave statistically in an exactly predictable manner, and a definite fraction will undergo radioactive decay in a given time. The rate of decay by this random process is given in terms of the *half-life,* the time required for half the nuclei of a radioactive isotope to decay. Analytical applications of radioactivity depend on the predictability of the rate of nuclear decay when large numbers of atoms are considered. Radiochemical methods are so sensitive that the radiation emitted by even a few atoms of radioactive materials can be detected and measured in a variety of ways. The important factor is that the number of disintegrations measured per unit time is directly proportional to the total number of atoms of a given radioactive isotope.

Most of the problems met in radiochemical analysis are those of any analytical procedure. Separations may incorporate operations such as precipitation, ion exchange, liquid–liquid extraction, and distillation. After separation the activity, and therefore the amount of radioactive species present, is determined by counting in an instrument such as a Geiger counter.

Uranium-238, the principal isotope of naturally occurring[1] uranium (99.3% ^{238}U, 0.7% ^{235}U), is radioactive and decays to thorium-234:

$$^{238}_{92}U \rightarrow \, ^{234}_{90}Th + \, ^{4}_{2}He \tag{10–9}$$

A helium nucleus, called an alpha particle, is produced in the process. Uranium-238 has a half-life of 4.498×10^9 years. Thorium-234, also radioactive, decays to protactinium-234:

$$^{234}_{90}Th \rightarrow \beta^- + \, ^{234}_{91}Pa \tag{10–10}$$

Here an electron, called a beta particle, is produced in addition to protactinium-234. The half-life of thorium-234 is 24.10 days. Protactinium, in turn, has a half-life of 1.18 min; it decays to uranium-234:

$$^{234}_{91}Pa \rightarrow \beta^- + \, ^{234}_{92}U \tag{10–11}$$

Although uranium-234 also is radioactive, its half-life is 248,000 years, so further decay is extremely slow. Uranium-234 decays to thorium-230, which has a half-life of 80,000 years.

Uranium-238 is normally in secular equilibrium with thorium-234. *Secular equilibrium* is the limiting case of radioactive equilibrium in which the half-life of a parent (here uranium-238) is many times greater than that of a daughter (here thorium-234). Under such conditions the radioactivity of the parent shows no appreciable change during a period

---

[1] Some commercially available uranium has different isotopic composition because of removal of uranium-235 from the natural material for nuclear applications.

equal to many half-lives of the daughter. After a period of at least 10 half-lives of the daughter, the relation

$$N_1 T_2 = N_2 T_1 \qquad\qquad (10\text{--}12)$$

applies, where N_1 is the number of atoms of the parent element that has a half-life T_1, and where N_2 is the number of atoms of the daughter that has a half-life T_2. Secular equilibrium of thorium-234 with uranium-238 is established to within 0.1% in about 8 months. In turn, secular equilibrium of protactinium-234 with thorium-234 is established in about 12 min.

Measurement of Radioactivity

Radioactivity refers to the ways in which a nucleus of an atom can release energy spontaneously (decay). Three common ways in which nuclei decay are emission of alpha (α) particles, beta (β) particles, and gamma (γ) rays. A variety of methods are employed to detect and measure these emissions. An important instrument for measuring beta particles is the Geiger counter. This device typically consists of a tube containing two electrodes and filled with a mixture of argon and an organic compound such as ethanol (Figure 10-4). One end is covered with a thin mica window to allow low-energy beta radiation to pass into the tube. From 1000 to 1500 V is applied across the two electrodes. An entering beta particle ionizes several molecules of the gas inside the tube. The generated electrons are drawn to the positive electrode with high velocity and attain sufficient energy to cause further ionization along the way. The resulting shower of electrons cascades to the positive electrode and causes a pulse of current through the circuit. This pulse trips a counting device and is recorded. After the pulse the effective voltage drops, ionization ceases, and the tube again is ready to count. The interval from the time a beta particle enters the tube until the system is ready for another count is called the dead time, which lasts about 200 μsec. Another beta particle entering during this time is not counted.

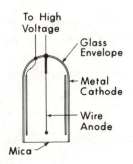

FIGURE 10–4. Schematic diagram of a Geiger tube.

Although more sophisticated apparatus (including proportional and scintillation counters with multichannel analyzers) has been developed, the Geiger counter remains the most important for beta counting because of its simplicity, ruggedness, and dependability.

Counting Thorium-234 Radioactivity

This analysis might be performed by determining the absolute disintegration rate for the thorium-234. Then no standard would be needed. Absolute counting requires much effort, however; ordinarily a comparison of the activities of a sample with a standard is far easier. The observed count rate differs from the absolute rate for the following reasons:

1. Some devices for measuring radioactivity can count two disintegrations separately if they occur at or nearly at the same time. A Geiger counter, however, counts such a dual event as one, because after each count there is a short interval during which the instrument does not respond. Corrections for simultaneous decay so as to give a more accurate count rate can be made by use of statistics and the laws of probability.
2. Activity from sources other than the sample may be counted. Cosmic rays and natural and artificial radioactivity in the surroundings contribute to the count rate. This background count must be measured separately and subtracted from the total.
3. Positioning of a sample under the counter tube is critical. Thorium-234 emits beta rays randomly in all directions. With a sample beneath the end window of a Geiger tube (the normal position) less than 50% of the radiation is intercepted.
4. Elastic collisions of beta particles with other molecules can result in a change of direction. These changes are classified as back scattering, fore scattering, and side scattering.
5. The mass of a sample itself absorbs some of the beta rays before they leave it. Sample self-absorption is minimized by keeping the mass of the sample small.

The overall decrease in count rate owing to losses from precipitation, geometry, scattering, and self-absorption can be calculated with the knowledge that 750 betas are emitted per minute per milligram of uranium from thorium-234 in secular equilibrium with aged uranium. Actually, beta rays from thorium-234 are of such low energy that they cannot penetrate the end window of a Geiger tube. The count is due to high-energy betas from protactinium-234 (half-life 1.18 min), which is in secular equilibrium with thorium-234; 750 of these also are emitted per minute per milligram of uranium.

In the following procedure, quantitative results depend on proper use of a standard. For reliable analyses, preparation and counting conditions for both samples and standards should be reproduced faithfully. Because radioactive decay is a random process, the indeterminate aspects of the data can be treated in a statistically ideal way.

Reliable results can be expected only if the separation of thorium-234 in the standards and sample is carried out simultaneously. Plan your operations before coming to the laboratory so that the precipitation and filtration steps can be completed in one period. Make a particular effort to handle each sample in the same manner, thus minimizing errors from variations in procedure.

Procedure (median time 3.7 hr)

Analytical Procedure

A standard thorium-234 solution may be provided in the form of an aged uranium solution of specified concentration.

Obtain a thorium sample and transfer it to a 100-ml volumetric flask, using 1 ml of 6 M HNO$_3$ during the washing stage to more completely transfer the thorium-234 from the sample tube.[2] Dilute to volume. Pipet

---

[2] The adsorption of thorium on the walls of the sample tube is decreased through the use of a dilute HNO$_3$ solution.

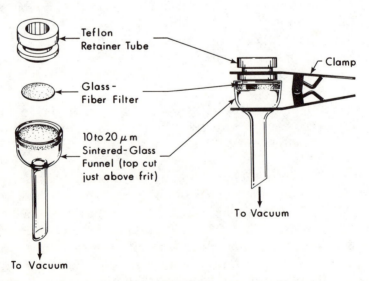

Teflon
Retainer Tube

Clamp

Glass-
Fiber Filter

10 to 20 μm
Sintered-Glass
Funnel (top cut
just above frit)

To Vacuum

To Vacuum

FIGURE 10–5. Sintered-glass filter assembly for collection of radioactive precipitates.

10-ml portions into 50-ml beakers. Pipet 10-ml portions of the standard thorium solution directly into another set of 50-ml beakers. Add 3 drops of 0.03 M $FeCl_3$ solution to each portion. Warm the solutions on a hot plate to about 60°C and add 0.5 M $(NH_4)_2CO_3$ dropwise, several drops at a time, to precipitate $Fe(OH)_3$ and $(NH_4)_2U_2O_7$. Continue the dropwise addition, swirling the solution vigorously, until the yellow precipitate, $(NH_4)_2U_2O_7$, redissolves, and then add about 1 ml excess. Carry all samples through each stage at the same time. Do not use stirring rods.[3] At the completion of this stage only a small amount of red-brown $Fe(OH)_3$ should remain. Leave the beakers and contents at 60°C for 4 or 5 min to coagulate the precipitate.

Tilt the beakers at an angle of about 30° so that the precipitate settles to the lip side, and allow to cool for 15 to 30 min. Record the hour and date.[4] Do not stir the solutions. When ready to filter, center a glass-fiber filter mat on a 10- to 20-μm sintered-glass filter (Figure 10-5). Clamp a retainer tube in the center of the filter mat and apply vacuum. Pour the liquid from the first beaker into the retainer. Try not to disturb the precipitate until as much as possible of the supernatant liquid has been filtered and then transfer it in one motion. Rinse the beaker and retainer tube with a few 1-ml portions of 0.1 M $(NH_4)_2CO_3$, using a plastic wash bottle with a delivery tube that points upward. Transfer the filter mat to a counting planchet with the aid of a spatula, without touching the $Fe(OH)_3$. The precipitate should be face up and the mat should lie flat. Allow the precipitate to air dry. When the filtration operations have been completed, remove residual $(NH_4)_2CO_3$ from the sintered-glass funnel by drawing a milliliter or two of 6 M HCl through the funnel. Finally, wash the precipitate with water.

Counting Procedure

Follow the directions posted on the Geiger counter for details of its operation.

Place an empty planchet in the topmost position beneath the Geiger tube. (*Care!* The bottom end of the tube is fragile.) Count for 3 min. Then count in order for 3 min each: first standard, first sample, second standard, second sample, third standard, through the set.[5] Repeat the count for each and then count the blank again for 3 min.

---

[3] The additional surface provided by stirring rods should be avoided; swirling the beaker contents provides sufficient mixing.

[4] The time of separation of thorium from uranium must be recorded because thorium-234 is both continually decaying and continually being formed from the parent uranium. Noting the time of separation makes it possible to allow for changes in concentration of thorium brought about by decay.

[5] Reproducibility of sample placement is important. Position each filter mat in the planchet so that the precipitate is centered under the tube of the Geiger counter.

The amount of thorium-234 present in a sample is directly proportional to the net count rate. No correction for decay is needed at this stage since the half-life is 24.1 days and both standards and samples were separated and are counted at the same time.

Calculate and report the picograms of thorium in the sample. Also report the date and time of the separation so that allowance can be made in grading for decay in the sample.

Calculations

Weight of Thorium-234 in the Sample

First calculate the weight of thorium in the standard in picograms per liter (1 g = 10^{12} picograms) as follows. For secular equilibrium

$$N_{Th} = T_{Th} N_U/T_U \tag{10-13}$$

where N_{Th} and N_U are the number of atoms of thorium-234 and uranium-238 present, and where T_{Th} and T_U are the half-lives of the two isotopes, 24.10 days and 4.498×10^9 years.

Since the number of atoms present is proportional to the number of moles, the concentration of thorium-234 M_{Th} in moles per liter is given by

$$M_{Th} = \left[\frac{24.10 \text{ days}}{(4.498 \times 10^9 \text{ yr}) (365.2 \text{ days/yr})} \right] M_U$$

$$= (1.467 \times 10^{-11}) (M_U) \tag{10-14}$$

Here M_U, the moles of uranium-238 present per liter of standard, may be expressed by

$$M_U = \frac{(\text{g U/ml std}) (0.993) (1000 \text{ ml/l})}{(\text{at. wt U})}$$

$$= \frac{(\text{g U/ml}) (0.993) (1000)}{(238.0)} \tag{10-15}$$

The 0.993 is included because only 99.3% of natural uranium occurs as uranium-238. Therefore,

$$M_{Th} = \frac{(1.467 \times 10^{-8}) (\text{g U/ml}) (0.993)}{238.0} \tag{10-16}$$

and the weight of thorium-234 in the standard in picograms per liter is

$\text{wt Th}_{std} =$

$$\frac{(1.467 \times 10^{-8})\,(0.993)\,(\text{g U/ml})\,(234.1 \text{ g Th/mole})\,(10^{12} \text{ pg/g})}{238.0}$$

$$(10\text{--}17)$$

Next, calculate the weight of thorium-234 in the sample. The relation between the weights of standard and sample and the corresponding counts is linear, so

$$\frac{\text{wt Th}_{smp}}{\text{wt Th}_{std}} = \frac{C_{smp}}{C_{std}} \qquad (10\text{--}18)$$

where C_{smp} and C_{std} are the net sample and net standard counts. Since the same volume, V_{10}, of both sample and standard was taken, the total weight in the initial 100 ml, or 0.1 liter, in picograms is

$\text{wt Th}_{smp} =$

$$\frac{(C_{smp})\,(0.1)\,(1.467 \times 10^{-8})\,(0.993)\,(\text{g U/ml})\,(234.1 \times 10^{12})}{(C_{std})\,(238.0)}$$

$$(10\text{--}19)$$

Statistical Calculations

Calculate the standard error of the average counts per minute and the 95% confidence limits that would be expected on the basis of the indeterminate errors associated with the radioactivity-measurement part of the experiment only. Also calculate the overall efficiency of the process, including losses due to separation, counting geometry, and radioactive decay. Assume that the beta emission rate from protactinium-234 in secular equilibrium with 1 mg of uranium is 750 per min. Calculate the overall efficiency from the relation

$$\text{overall efficiency in \%} = \frac{(\text{cpm for std})\,(100)}{(\text{mg U})\,(750)} \qquad (10\text{--}20)$$

where cpm is counts per minute. Calculate also the efficiency of separation and counting (corrected efficiency) by correcting the factor 750 to account for radioactive decay occurring between the time of separation and the time of counting. This correction is determined from the relation

$$\log N_0 = \log N_T + 0.301 \, (T/T_{\frac{1}{2}}) \tag{10-21}$$

where T is the elapsed time in days between filtration and counting, $T_{\frac{1}{2}}$ is 24.1 days, N_T is cpm for the standard, and N_0 is the corrected count rate. Corrected efficiency would be the same as overall efficiency only if counting were done immediately after precipitation.

$$\text{Corrected efficiency in } \% = \frac{(N_0) \, (100)}{(\text{mg U}) \, (750)} \tag{10-22}$$

To calculate the standard error of the average counts per minute and the 95% confidence limits for the weight of thorium calculated from Equation (10-19), using as a basis only the indeterminate errors associated with counting the radioactivity, apply the following (see also Example 10-1):

a.   The standard deviation of a single radiochemical measurement is equal to the square root of the total number of counts observed:

$$s = \sqrt{\text{Counts}}$$

b.   The average value of N measurements is more reliable than a single measurement by a factor of $\sqrt{N}$:

$$s = \sqrt{\text{Average counts}/N}$$

c.   The count rate (counts per unit time) and its standard deviation are obtained by dividing the total count and its standard deviation by the time of counting.

d.   The standard deviation of the sum or difference of two numbers is equal to the square root of the sum of the squares of the individual standard deviations.

e.   The relative standard deviation of the product or the quotient of two numbers is equal to the square root of the sum of the squares of the individual relative deviations.

Example 10-1.

Results for a set of radiochemical measurements. (Each count is made for 3 min. The letters in parentheses refer to the corresponding statement above.) Standard deviation for net

	Blank Count	Standard Number	Standard Count	Sample Number	Sample Count
	306	1	10,261	1	7,963
		2	10,536	2	lost
		3	10,176	3	8,013
		4	10,350	4	8,140
		1	10,183	1	7,985
		2	10,518	2	–
		3	10,301	3	8,110
		4	10,444	4	8,119
	252				
Total	558		82,769		48,330
Average count	279 ± 12 (b)		10,346 ± 36 (b)		8,055 ± 37 (b)
Average cpm	93 ± 4 (c)		3,449 ± 12 (c)		2,685 ± 12 (c)
Net cpm	–		3,356 ± 13 (d)		2,592 ± 13 (d)

$cpm = \sqrt{(12)^2 + (4)^2} = 13$ (d). Relative standard deviation in the ratio (net count rate for sample)/(net count rate for standard) equals the relative standard deviation in the weight of thorium.

$$\text{Relative standard deviation} = \sqrt{\left(\frac{13}{3356}\right)^2 + \left(\frac{13}{2592}\right)^2}$$

$$= 0.0068 \quad (e)$$

Absolute standard deviation = ± (0.0068) (wt of thorium). The 95% confidence limits are estimated to be two times the absolute standard deviation.

PROBLEMS

10–9.   A solution containing thorium-234 analyzed by the procedure of this experiment gave the following data: average counts for 3 min for blank, standard, and sample were 291, 9651, and 11,423; the standard thorium-234 solution contained 2.147 mg of uranium per milliliter. How much thorium-234 was present in the entire sample solution?

10–10.   Calculate the weight of protactinium-234 in secular equilibrium with 2.00 g of uranium.

10–11.   Calculate the percentage of uranium-234 in natural uranium from the half-lives and the percentage of uranium-238 present.

10–12.   In Experiment 10-5 the radiation counted is the beta emission from the reaction

$$^{234}Pa \rightarrow \beta^- + {}^{234}U \qquad \text{(half-life} = 1.18 \text{ min)}$$

Why is this activity detectable even after several weeks?

10–13.   If a pure sample of protactinium-234 were isolated, what fraction of it would remain after 10 min?

10–14.   A phosphate fertilizer containing phosphorus-32 as a radioactive tracer has a count rate of 800 cpm/g. How long will it take for the count rate to become 100 cpm/g? The half-life of phosphorus-32 is 14.5 days.

10–15.   Artefacts found in a prehistoric Indian campsite in Saskatchewan had a carbon-14 disintegration rate of 5.24 cpm/g. If carbon-14 in equilibrium with the atmosphere has a specific disintegration rate of 15.3 cpm/g, how old are the artefacts?

SELECTED REFERENCES

G. Friedlander, J. W. Kennedy, and J. M. Miller, *Nuclear and Radiochemistry,* 2nd ed., Wiley, New York, 1964.
S. Sethi and R. S. Rai, *Z. Anal. Chem.* **252,** 5 (1970). Describes separation of thorium-234 by adsorption on involuble iodates.

Chapter 11

SEPARATIONS IN ANALYSIS: CHROMATOGRAPHY

11–1. GENERAL BACKGROUND

In Chapter 10 the discussion on applications of liquid–liquid extraction and precipitation separations points out that these techniques can be used efficiently only when the enrichment factor between the wanted and unwanted constituents of a sample is highly favorable; otherwise, time-consuming multiple extractions or precipitations must be carried out.

One way in which two components having a poor enrichment factor can be separated is to dissolve the mixture in a moving (mobile) phase, bring it into contact with a stationary phase, and allow many equilibrations between the phases to take place as the mobile phase passes by. Even though equilibrium is unlikely to be attained completely at any point in the system because of the continuous movement of the mobile phase, the system still can give separations equivalent to hundreds or thousands of batch equilibrations, often rapidly within short distances. This operation is called *chromatography*.

A *chromatogram* is a record of the response of a detector at the end of a chromatographic column as a function of time. It consists of a base line corresponding to emergence of pure mobile phase on which are imposed peaks corresponding to emergence of mobile phase plus sample components. The volume of mobile phase required to elute a particular component is known as its *retention volume*. The volume required to elute air or other nonretained material is termed the *column-dead-space volume*. The *partition ratio* is the ratio of amount of solute in the stationary phase to the amount in the mobile phase. Experimentally, it is the ratio of retention volume minus dead space to dead space. The *resolution* of two chromatographic peaks is defined as the difference in retention volumes divided by the average peak width.

One example of a stationary phase is a liquid, immiscible in the mobile phase, held in a thin film on a finely divided solid; another is a solid onto which components of the mobile phase can be adsorbed; still another is a matrix of stationary ionic sites that exchange ions with the mobile phase. The stationary phase may be a coated solid in a tube, as in gas–liquid chromatography, or a sheet of fibres, as in paper chromatography. The separated components may be identified while still on the stationary phase, as in paper or thin-layer chromatography, or measured as they are removed (eluted) from the end of the stationary phase, as in gas–liquid or liquid–liquid chromatography.

In this chapter two types of chromatographic separations, ion exchange and gas–liquid, are illustrated. An additional experiment illustrates use of an ion-exchange resin to determine the total salt content of a sample; it is included to introduce the concepts and experimental techniques involved in working with columns as well as to illustrate a useful analytical method.

Ion-Exchange Resins

Ion-exchange resins are inert, insoluble materials containing acidic or basic sites that hold ions by electrostatic attraction. Most commercial ion exchangers are organic polymers; a widely used type is a copolymer of styrene and divinylbenzene. The degree of cross-linking between poly-

Styrene   Divinyl Benzene   Cross-linked Polystyrene

styrene chains is determined by the amount of divinylbenzene.[1] High cross-linking produces a more rigid resin that is less inclined to shrink or

---

[1] Dow Chemical Co. and Rohm & Haas Co. are two major producers of ion-exchange materials. Dow designates the percentage of divinylbenzene present in each resin by an "x" and a number immediately after the identifying number of the resin (8% is typical).

TABLE 11–1. SOME COMMERCIAL ION-EXCHANGE RESINS

Type	Exchange Site	Rohm & Haas	Dow
Strong acid	$-SO_3^-$	Amberlite IR-120	Dowex 50
Weak acid	$-COO^-$	Amberlite IR-50	–
Strong base	$-N^+(CH_3)_3$	Amberlite IRA-400	Dowex 1
Strong base	$-N^+(CH_3)_2(C_2H_5OH)$	Amberlite IRA-410	Dowex 2
Weak base	Polyamine	Amberlite IR-45	Dowex 3

swell upon change of composition of the solution with which it is in contact.

Ion-exchange sites on resins consist of strong or weak acidic or basic groups. The most common are $-SO_3^-$, $-COO^-$, $-N^+R_3$, and $-N^+R_2H$, where R is an organic group such as $-CH_3$ or $-C_2H_5OH$. In cross-linked polystyrene resins these groups generally are attached to benzene rings. Several typical resins are listed in Table 11-1.

The selectivity of a resin for various ions is governed primarily by the charge on the ion and its hydrated radius. The lower the charge density, the lower is the resin affinity for an ion. Specific bonding effects or nonspherical shapes may, however, throw individual species out of sequence. In general, selectivity increases with increasing charge on an ion and with atomic number for ions in the same group in the periodic table. Table 11-2 shows the selectivities of two resins. Although the relative ion affinities vary by only one or two orders of magnitude, repeated equilibrations are possible through use of a resin packed in a column, and exchange of ions on the resin for ions in solution can be made quantitative. The extent to which exchange between ions occurs depends on the affinity of the resin for each ion and on the concentrations of the ions in solution and on the resin.

Because the ion-exchange process is reversible, it provides an economical and convenient way to soften water. A strong-cation resin in the sodium form is placed in a column, and hard water is passed through it. The resin affinity for divalent ions such as calcium and magnesium is greater than for sodium, and so they replace sodium ions on the resin. When the column is almost completely converted to the calcium–magnesium form, it is regenerated with a concentrated solution of sodium chloride. By mass action the high concentration of sodium ions in solution displaces calcium and magnesium ions from the resin, converting it back to the sodium form for reuse. Ion-exchange resins may be cycled in this way almost indefinitely. The exchange capacity of most resins is about 2 milliequivalents per milliliter of resin.

TABLE 11–2. AFFINITY OF RESINS FOR SELECTED IONS[a]

Ion	Hydrated Radius, nm[b]	Relative Affinity
Ag^+	0.34	8.5
Cs^+	0.33	3.2
Rb^+	0.33	3.2
K^+	0.33	2.9
NH_4^+	0.33	–
Na^+	0.36	2.0
H^+	0.28	1.3
Li^+	0.38	1.0
Ba^{++}	0.40	11.5
Pb^{++}	0.40	9.9
Sr^{++}	0.41	6.5
Ca^{++}	0.41	5.2
Mg^{++}	0.43	3.3
ClO_4^-	0.34	10.0
NO_3^-	0.34	3.0
$C_2H_3O_2^-$	–	18
I^-	0.33	18
Br^-	0.33	3.5
Cl^-	0.33	1.0
F^-	0.35	0.1
OH^-	–	0.5

[a] For sulfonic acid (cation exchanger) and quaternary ammonium (anion exchanger) resins with 8% cross-linking. Values are relative to $Li^+ = 1.00$ or $Cl^- = 1.00$.

[b] The present recommendation for expressing distance is in terms of meters or in 10^3 fractions or multiples thereof, rather than in units such as angstroms (1 nm = 10^{-9} m = 10 Å).

11–2. ION EXCHANGE: TOTAL SALT IN A MIXTURE

Background

In this experiment a cation-exchange resin in the acid form is used to determine the total milliequivalents of salt in a sample. When a salt solution is passed through a column of this resin, cations in solution exchange with hydrogen ions on the resin to produce an amount of strong acid equivalent to the salt initially present. This acid is washed out of the column (eluted), and the effluent (eluate) is titrated with standard sodium hydroxide. The amount of base required is a direct measure of the amount of salt present in the original solution. For titration of weak acids, solutions of sodium hydroxide must be kept free of carbon dioxide, or the carbonate produced will obscure indicator end points. Since in this experiment the acid being titrated is strong, and an indicator that changes color in a low pH range is used, precautions to keep the sodium hydroxide free of carbonate are unnecessary.

Solutions of sodium hydroxide must be standardized after preparation. For this experiment potassium chloride is used as an acid–base primary standard in an unusual way. A sample is weighed and dissolved, the solution passed through an ion-exchange column, and the resulting hydrochloric acid titrated. Because the sample and standard are treated alike, errors in technique tend to cancel each other, particularly if alternation is used.

The resin must be converted initially to the acid form. This conversion process, called regeneration, is accomplished by passing hydrochloric acid through the column. The reaction is

$$Na^+R^- + H^+ \rightleftarrows H^+R^- + Na^+ \qquad (11–1)$$

where R^- is an exchange site on the resin. Although the resin has a greater affinity for sodium than for hydrogen ions, use of a large volume of concentrated acid permits quantitative conversion of the resin to the acidic form. Before a sample is placed on a column, the excess acid must be washed from the column to prevent its being titrated along with the sample and thereby producing high results.

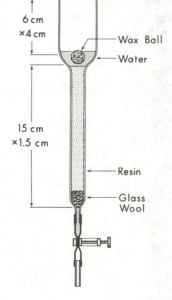

FIGURE 11–1. Ion-exchange column for analytical use.

Procedure (median time 4.2 hr)

Standardization of 0.1 M NaOH

Prepare about 1 liter of 0.1 M NaOH solution. Weigh accurately 0.3-g portions of pure, dry KCl into 50-ml beakers. Dissolve each portion in 1 to 2 ml of distilled water.

Obtain two 25-cm ion-exchange columns (Figure 11-1). Half fill the columns with water and insert a small portion of Pyrex glass wool.[1] Push the wool into a plug at the base of the column with a glass rod. Then add 50- to 100-mesh, strongly acidic cation resin until the resin level reaches the top of the narrow portion. Add a paraffin ball to minimize disturbance of the resin bed during addition of solutions. Regenerate the resin by pouring 100-ml portions of 3 M HCl through both columns. After regeneration, a column of the dimensions shown in Figure 11-1 has sufficient exchange capacity for about a dozen samples. Do not allow the level of the solution to fall below the surface of the resin; should it do so, backwash the column.[2] Pass distilled water through the column until the effluent is neutral. Test for neutrality by collecting about 30 ml of effluent and adding 1 drop of methyl red indicator. At neutrality, 1 drop of 0.1 M NaOH is sufficient to change the indicator to the alkaline color.

Add the KCl solution to the column in as small a volume as possible. Allow the solution to drain to near the top of the resin, and then with 3- to 5-ml portions of wash water rinse the beaker and the sides of the column. Allow each portion to drain to the level of the resin before adding the next. Repeat several times. Continue washing with water at a flow rate of about 5 ml/min (1 drop/sec) until the effluent is neutral; this will require 75 to 100 ml. Test a 10- to 20-ml portion collected from the column for neutrality with methyl red and NaOH as above. When the effluent is neutral, add several drops of methyl red indicator to the total effluent (including test portions), and titrate with 0.1 M NaOH solution. Calculate the molarity of the NaOH. Include in the titration volume the volume of NaOH used to test the sample effluent for acidity.

Analytical Procedure

Accurately weigh 0.20- to 0.30-g portions of the dry sample. Dissolve each portion in water, pass through the ion-exchange column, and titrate

---

[1] Previously unused Pyrex glass-wool fiber often has a film on its surface that may be a source of contamination to receiving flasks. Rinsing the wool with carbon tetrachloride prior to use eliminates this problem.

[2] Air bubbles become trapped in the resin if the liquid level drops below the top of the resin bed. These cause channeling and incomplete contact between the two phases. To backwash, connect a funnel to the column outlet with a 2-ft length of tubing and pass a flow of water up the column from the bottom. Alternatively, connect a wash bottle with an upward-pointing tip to the column outlet and use it to force water up the column. The resin should be raised by the force of the water and then settle back to give a uniform bed.

with standard NaOH solution as above.[3] If the indicator fades, add more as needed. Report the salt content of the sample as milliequivalents of salt per gram of sample.

Return the columns filled with the cation-exchange resin. Solutions of NaOH attack glass, so do not allow them to stand overnight in a buret or for more than a few days in a glass storage bottle. Tightly stoppered polyethylene bottles are best for storage of NaOH solutions.

Calculations

The equivalent weight of a substance in ion exchange is the number of grams of the substance that will replace 1 mole of H^+ or OH^- on a resin. For example, since 1 mole of pure KCl will displace 1 mole of H^+, the equivalent weight of KCl is the same as the molecular weight, 74.56, and the milliequivalent weight (meq wt) is 0.07456. For K_2SO_4 the milliequivalent weight is mol wt/2000 or 0.08714 g/meq.

Milliequivalents of salt per gram of sample are reported here because the identity of the compound, and consequently the number of millimoles in the sample, are unknown.

PROBLEMS

11–1.   A 0.2512-g sample of a mixture of sodium salts required 40.98 ml of 0.0972 M sodium hydroxide for titration after passage through an ion-exchange column in the hydrogen form. How many milliequivalents were present in the sample? How many milliequivalents were present per gram?

11–2.   A sample weighing 0.2386 g containing iron(II) sulfate and ammonium sulfate was dissolved in 10 ml of water and passed through a cation-exchange column in the hydrogen form. The effluent required 26.62 ml of 0.1120 M sodium hydroxide. Calculate the percentage of SO_3 in the sample.

11–3.   What types of salts cannot be determined by the procedure of this experiment?

11–4.   A 0.2218-g mixture containing only sodium chloride and sodium nitrate required 36.71 ml of 0.0995 M sodium hydroxide for titration after treatment by the procedure of this

---

[3] Errors in this experiment are caused mostly by nonquantitative transfer of samples, by incomplete elution of samples, and by failure to use alternation.

experiment. What was the percentage of each salt in the mixture?

11–5.† Give the number of milliequivalents per gram of each of the following that would be expected after passage through a cation-exchange column in the hydrogen form and titration of the effluent with sodium hydroxide (methyl red indicator): (a) KNO_3 ; (b) $NaOH$; (c) $LiBr$; (d) Na_3PO_4 ; (e) $NaCN$; (f) $NaC_2H_3O_2$.

SELECTED REFERENCES

W. M. MacNevin, M. G. Riley, and T. R. Sweet, *J. Chem. Educ.* 28, 389 (1951).
W. Rieman and H. F. Walton, *Ion Exchange in Analytical Chemistry*, Pergamon, New York, 1970.

11–3. ION-EXCHANGE SEPARATION AND COMPLEXATION TITRATIONS: DETERMINATION OF ZINC AND NICKEL

Background

In Section 11-2, ion exchange is shown to be useful for quantitative replacement of one cation by another, and it is pointed out that this method of cation or anion replacement has a variety of analytical applications. This experiment illustrates another application of ion exchange, the separation of two components of a mixture by formation of a complex. In 2 M hydrochloric acid, zinc(II) forms the negatively charged complexes $ZnCl_3^-$ and $ZnCl_4^=$ that can exchange with chloride ions on an anion-exchange resin in the chloride form. The equilibrium lies far on the side of the resin-zinc chloride complexes:

$$ZnCl_3^- + R^+Cl^- \rightleftarrows R^+ZnCl_3^- + Cl^- \qquad (11-2)$$

and

$$ZnCl_4^= + 2R^+Cl^- \rightleftarrows (R^+)_2 ZnCl_4^= + 2Cl^- \qquad (11-3)$$

Nickel(II) does not form a stable complex with chloride under these conditions, and since cations do not interact with anion-exchange sites, nickel is not held by the resin. If anion resin in the chloride form is placed in a column and a solution containing zinc and nickel in 2 M hydrochloric acid is passed through it, the zinc is held by the resin, but the nickel is not. Once the nickel has been eluted, distilled water is passed through the column. This lowers the chloride ion concentration to a level where the

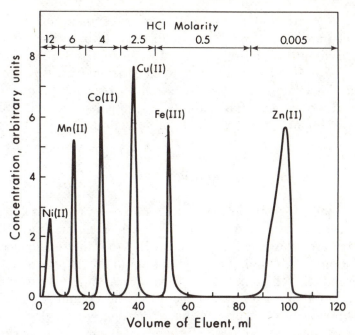

FIGURE 11–2. Separation of six transition metals on an anion-exchange resin by elution with successively lower concentrations of hydrochloric acid. (Adapted from Kraus and Moore; see reference at end of experiment.)

anionic zinc–chloro complexes are no longer stable; the zinc is then readily eluted from the resin. The chloro complexes of the transition metals vary in stability over a wide range, and it is possible, by systematically changing the concentration of hydrochloric acid eluent, to separate a number of them by the procedure described here (Figure 11-2).

After separation, the zinc and nickel may be determined in several ways. One convenient procedure is by titration with EDTA. Zinc may be titrated in a solution buffered at about pH 10 with ammonia; Calmagite is a suitable indicator for this purpose. Although nickel also may be titrated in ammoniacal solution at about the same pH, the nickel–Calmagite complex dissociates slowly; therefore the indicator murexide, the ammonium salt of purpuric acid, is recommended for nickel. Murexide, being unstable in solution, is best added as a solid. Because so little is required for a titration, it is diluted with an inert material such as sodium chloride.

Procedure (median time 5.5 hr)

Preparation of Standard 0.02 M EDTA

Place about 4 g of reagent-grade disodium ethylenediaminetetraacetate dihydrate (EDTA) in a weighing bottle, and dry in an oven at

80°C for 2 hr.[1] After it has cooled to room temperature, weigh to the nearest milligram about 3.7 g of the dry salt into a clean, dry 400-ml beaker. Add about 200 ml of deionized water to dissolve the solid; gentle warming at 50 to 60°C with occasional stirring increases the rate of dissolution. When the salt is dissolved, transfer the solution quantitatively to a 500-ml volumetric flask, dilute to volume, and mix. Store the solution in a 500-ml bottle, preferably of polyethylene.

Separation of Zinc and Nickel by Anion Exchange

Use deionized water throughout. Obtain a sample containing 3 to 6 millimoles each of zinc and nickel, dissolve if necessary, and dilute in a 100-ml volumetric flask, adding sufficient HCl that the final concentration of acid is about 2 M. Obtain two columns about 35 cm long and 1.3 cm in diameter, and fill to about the 25-cm level with 50- to 100-mesh, strong-base anion-exchange resin. Backwash the columns if necessary to remove air bubbles. Break up any lumps of resin with a glass rod. A wax ball on top of the resin reduces disturbance of the resin during addition of solution.[2] Wash the columns with about 50 ml of 6 M NH_3 to remove any metals present as anionic species, followed by 100 ml of deionized water and then 100 ml of 2 M HCl. The flow rate should be 4 to 5 ml/min. Do not allow the liquid level to drop below the surface of the resin, or channeling and air-bubble entrapment may result.

Once the columns are prepared, allow the liquid to fall to just above the resin, and then pipet a 10-ml portion of sample onto the top of each column. Collect the effluent in a 250-ml conical flask. Wash the inner wall of the column above the resin with several 3- to 4-ml portions of 2 M HCl, permitting the liquid level to come to the surface of the resin bed each time before adding the next portion. Then pass about 50 ml of 2 M HCl through the column to elute the nickel. After elution is complete, place the flask on a steam bath or hot plate in a hood and evaporate to dryness. Do not heat the residue too strongly, or the readily soluble nickel chloride may be converted to insoluble nickel oxide.

Place a clean 500-ml conical flask under the column outlet, and elute the zinc from the column by passing 50 to 100 ml of distilled water through the column at a flow rate of 3 to 4 ml/min. After the titration the column will be washed with an additional portion of water and the effluent tested to ensure complete removal of zinc.

---

[1] The dihydrate of reagent-grade Na_2H_2Y is sufficiently pure to be used as a standard at the level of a few parts per thousand. Excess moisture present in the salt is removed by gentle warming (80°C); heating at higher temperatures will cause loss of water of hydration and should be avoided.

[2] The diameter of the ball should be slightly less than the inner diameter of the column.

Titration of Zinc with EDTA

Remove the flask and add 10 ml of pH 10 buffer (prepared by dissolution of 10 g of NH_4Cl in 150 ml of 6 M NH_3) and 5 drops of Calmagite indicator.[3] Titrate with 0.02 M EDTA solution to the point where one or two drops of titrant gives the greatest color change between the wine-red of the zinc–Calmagite complex and the blue of the free indicator. At the end point, record the volume of titrant and pass another 50 ml of distilled water through the column, collecting it in the titrated solution. If the wine-red color of the indicator returns, additional zinc has been removed from the column. In this case, simply continue the titration to the blue end point. Record the new volume of titrant, and repeat the washing and EDTA titration until all the zinc has been removed from the column, as shown by no color change on further washing. Record the total volume of EDTA solution required.

Calculate and report the total grams of zinc present in the original sample.

Titration of Nickel with EDTA

Dissolve the residue of nickel chloride in 25 ml of deionized water, add 10 ml of pH 10 buffer, and dilute the solution to about 100 ml with water. Add 0.2 g of solid 0.2% murexide indicator, and titrate immediately with standard 0.02 M EDTA solution until the yellow of the nickel–murexide complex changes to the purple of the free indicator. Record the volume of EDTA solution required.

Calculate and report the total grams of nickel in the original sample.

PROBLEM

11–6.   A 1.2437-g sample of an alloy was dissolved and analyzed for zinc and nickel by the procedure of this experiment. If 3.621 g of $Na_2H_2Y\cdot 2H_2O$ was taken for preparation of the standard titrant, and 33.84 and 38.71 ml were required for titration of the zinc and the nickel eluted from the column, what was the percentage of each metal in the sample?

---

[3] The pH of the solution must be near 10 for optimum functioning of the Calmagite indicator (Experiment 3-3).

SELECTED REFERENCES

W. J. Blaedel and H. T. Knight, *Anal. Chem.* **26**, 741 (1954). A study of $Na_2H_2Y\cdot2H_2O$ as a standard for EDTA titrations.

K. A. Kraus and G. E. Moore, *J. Amer. Chem. Soc.* **75**, 1460 (1953). A description of the separation by anion-exchange chromatography of the transition metals as their chloro complexes.

11–4. ION-EXCHANGE CHROMATOGRAPHY: DETERMINATION OF CHLORIDE AND BROMIDE IN A MIXTURE

Background

The preceding two experiments show that ion exchange is useful for separating ions on the basis of differing ionic charge. Ion exchange is a more powerful tool for separations than this, however, and with a little refinement can be used to separate ions similar in size, charge, and chemical behavior. With highly refined technique even the separation of isotopes, such as sodium-22 from sodium-24, is possible. This experiment illustrates the use of an anion-exchange resin to separate two similar anions, chloride and bromide. The resin initially is converted to the nitrate form. When a solution containing chloride and bromide salts is added to the resin, the equilibria established are

$$R^+NO_3^- + Cl^- \rightleftarrows R^+Cl^- + NO_3^- \tag{11–4}$$

and

$$R^+NO_3^- + Br^- \rightleftarrows R^+Br^- + NO_3^- \tag{11–5}$$

The relative affinity of the resin for chloride ion is lower than for bromide ion (Table 11-2). Therefore, when the resin is eluted with a solution of a nitrate salt, chloride moves down the column more readily than bromide and appears in the effluent first. The amount of chloride and bromide in the sample can be determined by titrating portions of the effluent with standard silver nitrate solution. Separation is more nearly complete as the band of resin containing the initial sample is made narrower, as the flow rate is decreased, and as the nitrate concentration is decreased.

Procedure (median time 8.4 hr)

In this experiment, use only distilled water purified by passage through an anion or a mixed-bed ion-exchange column. Prepare 250 ml of standard 0.05 M $AgNO_3$ as follows. Weigh a beaker accurately on an analytical balance. Using a triple-beam balance, weigh approximately the needed solid $AgNO_3$ into the beaker, and weigh the beaker and contents accurately again. Dissolve the salt, and transfer to a 250-ml volumetric flask. (*Caution:* Do not spill $AgNO_3$ in the balance.)

Weigh 4.5 g of dry sample, dissolve, transfer to a 250-ml volumetric flask, and dilute to volume.

Prepare 1 liter of approximately 2 *M* $NaNO_3$ solution. By diluting a portion of this solution, prepare 1 liter of 0.4 *M* $NaNO_3$ solution. Obtain a siphon pipet (Figure 11-3) and a 35-cm column with a 22- by 1.3-cm bed of anion-exchange resin (50- to 100- mesh Dowex 2 X 8, for example). Backwash the column if necessary to remove air bubbles. Break up any lumps of resin with a glass rod. Mount a siphon pipet under the column to collect and transfer the effluent (Figure 11-3). Place a 1.2-cm wax ball on top of the resin to protect the resin bed from being disturbed when solution is added. Remove halide present on the resin by elution with 15-ml portions of 2 *M* $NaNO_3$ until a test of the effluent for halide is negative. Add 10 ml of 0.4 *M* $NaNO_3$ to the resin, and allow the level to fall to about 1 cm above the resin; repeat this operation once. During the preceding operations, measure the average volume delivered by the siphon pipet with a graduated cylinder. The volume delivered may vary

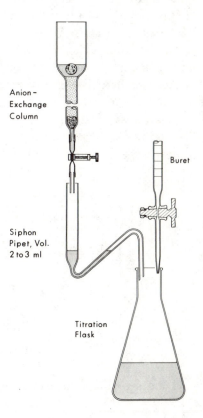

FIGURE 11–3. Assembly for titration of aliquots from ion-exchange column.

Anion – Exchange Column

Buret

Siphon Pipet, Vol. 2 to 3 ml

Titration Flask

appreciably from portion to portion. This variation is not critical, however, because the volume is used only to determine the general shape of the elution curve. The volume should be in the range 3 to 4 ml.

Allow the liquid level in the column to fall just to the top of the resin. Fill a buret with standard $AgNO_3$. Pipet a 10-ml aliquot of the sample solution onto the top of the column. Collect the effluent in a 500-ml flask containing about 1 g of $NaHCO_3$ and 5 ml of 5% K_2CrO_4. Wash the sides of the column with four 2- to 3-ml portions of 0.4 M $NaNO_3$. Each time bring the liquid level to the top of the resin before adding the next portion. Carefully add about 10 ml of 0.4 M $NaNO_3$ and elute at a rate of about 3 ml/min (about 1 drop/sec). If possible, do not interrupt the flow during elution. Keep a record of the number of siphon-pipet portions that are delivered. Titrate successive portions with $AgNO_3$ solution, starting with about the fourth portion. Record the buret reading and siphon-pipet number after titration of each portion. Add more $NaNO_3$ solution to the reservoir as needed. When a minimum in the volume of $AgNO_3$ required per portion is reached, change to 2 M $NaNO_3$ eluent and continue until all the bromide has been eluted.

Prepare the column for the next sample by passing about 20 ml of 0.4 M $NaNO_3$ through it. Once an elution curve has been obtained, in subsequent runs careful titration is necessary only of those portions in the region of the chloride–bromide separation point (the minimum between the peaks in Figure 11-4) and of the solution containing the completely eluted bromide. The end points are sharper if the first titration is delayed until just before the minimum in the elution curve; the effect of exposure of AgCl to light is thereby minimized. When finished, wash the resin free of $NaNO_3$ with two 50-ml portions of distilled water.

If the separation obtained is not satisfactory, another run at a lower flow rate or with a lower nitrate concentration will improve the separation somewhat, though elution will take longer.

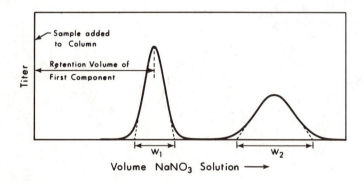

FIGURE 11–4. Idealized and smoothed elution plot of chloride–bromide separation by ion-exchange chromatography.

Determine indicator blanks by first adding 5 ml of 5% K_2CrO_4 and 1 g of $CaCO_3$ to about the same volume and concentration of $NaNO_3$ solution that was present at each end point and then titrating each with silver nitrate.

Report the percentage of chloride and, on a second report form, the percentage of bromide in the sample. The percentage of chloride is given by

$$\% \; Cl = \frac{(ml_{AgNO_3}) \, (M_{AgNO_3}) \, (at. \; wt_{Cl}) \, (250) \, (100)}{(1000) \, (wt_{smp}) \, (V_{10})} \qquad (11\text{–}6)$$

where ml_{AgNO_3} is the volume of $AgNO_3$ required to titrate to the chloride–bromide separation point and V_{10} is the volume of the calibrated 10-ml pipet.

Also construct a graph of $AgNO_3$ titer against siphon-pipet portion number. Calculate the number of theoretical plates for the chloride peak from the expression $16(V/W)^2$, where V is the retention volume (the volume from sample addition to the peak maximum) and W is the peak width. The peak width is the width at the base-line intercepts of the tangents to the inflection points (Figure 11-4; see also Experiment 11-5).

PROBLEMS

11–7.   A 1.7853-g sample containing soluble chloride and bromide salts was dissolved in 100.0 ml of water, and a 10.03-ml portion required 24.62 ml of a silver nitrate solution to the chloride–bromide separation point. The silver nitrate solution contained 2.1171 g in a 250-ml volumetric flask. The volume of silver nitrate required to titrate a 10.03-ml aliquot was 36.31 ml. What was the percentage of chloride and of bromide in the sample?

11–8.   Solutions of the following mixtures were put through a cation-exchange column in the hydrogen form: (a) $NaNO_3$ and $Ca(NO_3)_2$; (b) NaBr, KBr, and $CaBr_2$; (c) K_2SO_4, $MgSO_4$, and Ag_2SO_4. What was the order of appearance of ions in the effluent (Table 11-2) if the eluent was a dilute solution of hydrochloric acid?

11–9.   The effectiveness of a chromatographic separation depends on the location of the peak maxima upon elution (retention volume) and on the width of the peaks. List four factors that affect these two parameters, and indicate the direction and extent of each effect.

11–10.   The zinc and cadmium in a 0.2411-g sample are separated by cation-exchange chromatography. The effluent portions are titrated with 0.02467 M EDTA solution. The separation point occurs at 32.16 ml, and the total elution of both metals occurs at 41.67 ml. What is the percentage of each metal in the sample? (*Hint:* Predict which metal will be eluted first.)

SELECTED REFERENCES

W. Rieman and H. F. Walton, *Ion Exchange in Analytical Chemistry,* Pergamon, New York, 1970.
R. H. Schuler, A. C. Boyd, Jr., and D. J. Kay, *J. Chem. Educ.* **28,** 192 (1951).

11–5. GAS CHROMATOGRAPHY: SEPARATION AND ANALYSIS OF A MIXTURE OF BENZENE AND CYCLOHEXANE

Background

Quantitative organic analyses based on chemical reactions often involve slow rates and small equilibrium constants. For this reason, physical methods of organic analysis not requiring a chemical reaction generally are preferred. Of the many physical methods of carrying out organic analysis, such as nuclear magnetic resonance and mass, infrared, and ultraviolet spectroscopy, none is more important than gas chromatography. Since physical methods require instruments, the details of operations are not so directly under the control of the analyst as in wet chemical methods. He must rely on an instrument, and unless he understands how the instrument is intended to function, he will not be able to judge whether it is operating properly, what levels of precision and accuracy to expect, or how to prepare samples and standards to obtain optimum results.

There is a growing mythology that the scientist of the future will need only a finger for pushing buttons on black boxes, that the black boxes will do the rest, and that somehow the finger will need neither intelligence nor experimental competence behind it. Although appropriate instruments and machines undeniably enable us to do more and make our lives more interesting, the black box still requires an appropriate sample before it can deliver useful output. The preparation of the sample requires skill and care as well as knowledge of chemical reactions. In addition, knowledge of the principles and operation of the black boxes is essential.

Gas Chromatography

Gas chromatography (Figure 11-5) is a technique for carrying out the separation and measurement of mixtures of materials that can be volatilized. These materials may be gases, liquids, or solids that have appreciable vapor pressures at temperatures up to a few hundred degrees. In gas chromatography a stationary phase, generally a stable nonvolatile liquid, is spread in a thin film on a solid support. The support may be a porous granular solid or the inner wall of a capillary. A carrier gas acts as an inert moving phase to transport sample components from an injection point at the head of the column through the column to a detector. Sample injection is an arrangement by which a solid, liquid, or gaseous sample is transmitted as a short pulse into the carrier-gas stream before it enters the column. The sample should be vaporized and carried to the leading end of the column in negligible time. The detector, commonly a thermal conductivity cell, monitors the composition of the carrier-gas stream as it leaves the column. Simple, sensitive, and stable, the thermal conductivity cell has contributed in a major way to the explosive growth of gas chromatography. A significant advantage is that it provides a recorder response proportional to concentration of substance in the effluent from a column. Although peak areas are proportional to the amount of a given constituent, they are *not* the same for the same quantities of *different* substances. The composition of a mixture therefore must be calculated from measurements of peak-area ratios of the components of a known mixture and of the sample.

A critical aspect of gas–liquid chromatography is preparation of the

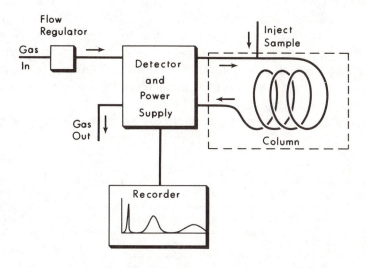

FIGURE 11—5.   Schematic drawing of a gas chromatograph.

column. The solid support must possess a uniform structure of high surface area, and it must be coated with a thin but even film of the stationary liquid phase. This mixture must then be packed carefully into a long tube so that channeling or plugging is minimal.

In this experiment a mixture of cyclohexane and benzene is separated and analyzed by gas chromatography on a column of di-n-decyl phthalate coated on Chromosorb P (a finely crushed special firebrick). The close boiling points of benzene and cyclohexane, 80.1 and 80.7°C, make separation by distillation difficult. The experiment illustrates the technique of using a standard mixture when the amount of sample cannot be measured precisely. In addition, it provides experience in both the handling of moderately volatile substances and the use of a recorder for the collection of data.

Procedure (median time 2.0 hr)

Prepare an accurately known mixture of cyclohexane and benzene by weighing about 3 g of each into a weighing bottle. Use a weighing technique that takes into account evaporation of the first component added.[1] Obtain the sample and a syringe.[2]

Immerse the tip of the syringe in the sample and withdraw more than is actually needed for an injection (2 to 10 μl). Invert the syringe, retract the plunger slightly to free the sample of air bubbles, and then return the plunger to the appropriate sample setting. With the syringe still inverted, retract the plunger to include a sample of air of approximately three to five times the volume of liquid (if capacity of syringe is adequate). Then insert the needle through the injection port and immediately push the plunger all the way down. Withdraw the needle. The insertion, injection, and withdrawal should be done in one smooth, continuous operation. Evaporate remaining traces of sample from the syringe by moving the plunger back and forth a few times. Time on the instrument can be saved by practicing the technique for injection on a spare column before attempting a run.

In normal chromatographic practice one sample is injected and the entire chromatogram obtained before the next is injected. Mark the

---

[1] One method is to weigh about 3 g of benzene to the nearest 0.1 mg into a clean, dry weighing bottle. Remove the cover for about the length of time required to add a portion of cyclohexane (a few seconds); re-cover and reweigh. Record this weight and quickly add about 3 g of cyclohexane (previously weighed approximately into a test tube or flask) to the bottle; re-cover and reweigh again. The total weight of benzene lost by evaporation is then twice the amount measured during the first removal of the cover.

[2] A syringe capable of accommodating about 10 μl of sample plus enough air to give a discernible recorder signal is required. A gas-chromatographic syringe of at least 2.5 μl capacity with a ½-in. needle is suitable.

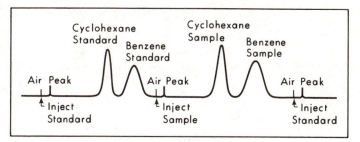

FIGURE 11—6. Chromatogram illustrating injection of standard and sample mixtures of benzene and cyclohexane.

injection points and mixtures injected on the chart to avoid confusion during interpretation (Figure 11-6).

If an injection is spoiled, or thought to be spoiled, wait for the length of time required to elute the benzene before injecting another sample. Since some sample usually enters the column even in poor injections, immediate injection of another sample may result in overlapping peaks. Do not inject a second sample even if no air peak is observed. The amount of air injected is somewhat independent of the amount of sample that is injected, so the absence of an air peak is not a reliable indication of absence of sample.

Determine peak areas by height-width measurements as outlined in Experiment 9-2. Calculate and report the weight percent of cyclohexane in the sample. Cyclohexane has a shorter retention time than benzene.

Instructions for Operation of a Gas Chromatograph[3]

1. Adjust the pressure of the helium cylinder at the second stage of the reducing valve to read 17 to 20 pounds/sq in. above atmospheric pressure (psig). With a small stopcock attached to the second stage, adjust the flow of carrier gas through the apparatus so that a steady stream of bubbles is produced when a small beaker of water is held under the outlet tube (about 50 ml/min).

2. Turn on the power-supply control unit. Set the filament current to about 130 milliamperes.

3. Warm up the recorder (if a tube model) by turning the lower right knob to standby a few minutes before use. Lower the recorder pen and see that it is writing properly. If the pen will not move to the right side of the chart, check the gas flow again. Notify the laboratory instructor if difficulty is encountered. The base line should be positioned 1 to 2 cm from the right edge of the paper.

---

[3] This procedure applies to an apparatus consisting of a Gow-Mac thermal conductivity detector (Model 92851) and power supply (Model 9999C) and a Sargent recorder (Model SR).

4.   When ready to make a run, turn the recorder knob to drive, set the pen on the paper, and inject a sample.

5.   When finished, turn off the recorder, power-supply control unit, and helium flow, in that order, to avoid damage to the thermal conductivity cell.

Calculations

The composition of a sample in weight percent can be calculated from the standard and sample peak-area ratios and the known composition of the standard. In this calculation, allowance must be made for the fact that all substances do not cause the same size of detector signal per unit weight of material. Assuming linearity between detector signal and amount of material, use the following formula to calculate weight percent of one component a in the mixture:

$$P_{ax} = \frac{W_a(r_x/r_s)(100)}{W_a(r_x/r_s) + W_b} \qquad (11-7)$$

where P denotes weight percent, W is the weight of a or b in the standard, r is the ratio of the peak area of the first component eluted to the area of the second component eluted, subscripts a and b are the first and second components, and s and x refer to standard and sample.

Several parameters of the separation system can be obtained from the chromatograms. The following calculations are suggested:

1.   Calculate the number of theroretical plates for a cyclohexane and a benzene peak. The number of theoretical plates is given by $16(V/W)^2$, where V is the retention volume and W the peak width at the base line (Figure 11-4).

2.   Calculate the resolution R of the two peaks. Resolution is equal to the difference in retention volumes of two peaks divided by their average peak width $(V_2 - V_1)/0.5(W_2 + W_1)$.

3.   Calculate the partition ratio for each component. The partition ratio is the ratio of net retention volume to dead-space volume $(V - V_{ds})/V_{ds}$.

PROBLEMS

11-11.   A sample containing a mixture of 1-chlorobutane and 1-chloropentane was separated and determined by the procedure of this experiment. A standard mixture was prepared to contain

2.643 g of 1-chlorobutane and 3.006 g of 1-chloropentane. For a standard run the first peak had an area of 21.41 cm² and the second an area of 27.41 cm². For a sample run the first peak had an area of 18.36 cm² and the second an area of 17.96 cm². What was the percentage of 1-chlorobutane in the sample?

11–12. A mixture of butane and pentane was injected into a gas-chromatographic column with the following results: Air peak maximum appeared after 0.70 min with peak base width equivalent to 0.20 min; butane maximum appeared after 3.5 min with base width equivalent to 0.80 min; pentane maximum appeared after 4.8 min with base width equivalent to 0.90 min. What is the partition ratio for butane on this column? What is the resolution of butane and pentane?

11–13. Calculate the resolution of the peaks shown in Figure 11-4. Calculate also the number of theoretical plates for each peak.

11–14. A compound is not eluted from a column in a reasonable time. Independent measurements indicate that its partition ratio is about 100,000. What changes in experimental conditions could be recommended to decrease the retention time?

11–15.† In a given gas-chromatographic experiment, adjacent hydrocarbon homologs were found to have a resolution of 9.5. From the definition of resolution, estimate the maximum number of separate components that could be isolated in reasonable purity, under the same experimental conditions, from a complex hydrocarbon mixture containing compounds ranging from pentane to octadecane. (*Hint:* Assume that a resolution of 1 is adequate for the isolation of one substance from another.)

SELECTED REFERENCES

W. E. Harris and H. W. Habgood, *Programmed Temperature Gas Chromatography,* Wiley, New York, 1966, Chapter 1 and p. 242. Covers general method. Includes a description of the apparatus used in this experiment.

H. A. Laitinen and W. E. Harris, *Chemical Analysis,* 2nd ed., McGraw-Hill, New York, 1974, Chapters 24 and 25.

A. B. Littlewood, *Gas Chromatography,* 2nd ed., Academic Press, New York, 1970.

Chapter 12

ANALYTICAL PROBLEM SOLVING AND USE OF THE LITERATURE

12–1. ANALYTICAL PROBLEM SOLVING

Not the fact avails, but the use you make of it.

Ralph Waldo Emerson

Background

One of the fundamental challenges of chemistry lies in investigating what and how much of a substance is present in a sample. Questions of this kind are met constantly by chemists in all fields—the organic or inorganic chemist who is characterizing new products, the physical chemist who is studying the kinetics of a system, and the industrial chemist who is trying to improve yield and quality of a commercial product, as well as the analytical chemist who is developing and using new methods. Similar challenges are met by scientists in many other fields, including biology, medicine, engineering, and agriculture, who may seek information about pollution in the air, water, and land around us, about insecticide and herbicide concentrations in food, and about the composition of the moon.

Up to this point involvement in details of individual experiments may have obscured one of the basic purposes of analytical chemistry—the solution of chemical problems. The analytical chemist is a solver of problems. To give some perspective on applications of analytical thinking, this section poses some analytical problems. The selection of an appropriate method of analysis depends on a clear statement of the problem in terms of the nature of the substance to be determined, its approximate concentration, and the nature of other materials present.

248

Example 12–1.

In a mixture of hydrochloric and nitric acids, how can the amount of each be determined if no other solutes are present? Several methods could be considered. The total of the hydrochloric plus the nitric acid could be obtained by titration with standard sodium hydroxide. The amount of hydrochloric acid could be obtained by precipitation and weighing of chloride as silver chloride. The amount of nitric acid would be obtained by difference.

PROBLEMS

Although in some instances other methods might be preferable, each of the following problems can be solved by application of the methods described in this text. Describe any necessary modifications.

12–1. Suggest methods for determining the following, assuming each is present in concentrations of at least several percent:

(a) the hydrochloric acid and sodium chloride in a mixture of the two;

(b) the hydrochloric acid and sulfuric acid in a mixture of the two;

(c) the nitric acid and sodium nitrate in a mixture of the two;

(d) the bicarbonate in a sodium chloride brine;

(e) the ethylene glycol in an automobile cooling system;

(f) the iron(III) and iron(II) in a solution containing salts of these two ions;

(g) the dichromate and chromium(III) in a solution containing chromium(III) sulfate, potassium dichromate, and potassium sulfate;

(h) the copper and nickel in an alloy containing iron and tin;

(i) the permanganate and periodate in a mixture of the two potassium salts.

12–2. Devise a procedure for determining the amount of each component in the following mixtures, assuming each to contain a few percent of inert material: (a) hydrochloric acid and ammonium chloride; (b) calcium sulfate and sodium sulfate; (c) iron(III) nitrate and aluminum nitrate.

12–3. Suggest a method for determining the amount of iron in: (a) a rock consisting mostly of the mineral iron pyrite, FeS_2; (b) wine.

12–4. How could the concentration of cyanide in an approximately 0.1 M solution of the sodium salt be determined?

12–5. A worker in a chemical plant mistakenly added an unknown quantity of iron(III) chloride to a mixing chamber containing the chloride salts of aluminum, calcium, sodium, and iron(II). How could he determine how much iron(III) was present in the final mixture?

12–6. An organic chemist needs to know whether a bottle of ethylene glycol he plans to use in a synthesis is contaminated with glycerol, $CH_2OHCHOHCH_2OH$. How could he determine to the nearest percent the amount of glycerol present?

12–7. Maintenance of jet aircraft engines includes careful monitoring of bearings for excessive wear. An engineer wishes to take a sample of lubricating oil from a jet engine and determine whether the metal content of the oil may be used as a measure of the level of bearing wear. What methods could he use for a rapid analysis of iron, copper, cobalt, and nickel at the 1% relative-error level in the oil?

12–8. An organic research chemist in a pharmaceutical company has just isolated a new compound from juniper berries that he hopes will neutralize the physiological action of a hallucinogenic drug. The compound melts at 72°C, boils without decomposition at 180°C, and contains both carboxylic acid (–COOH) and primary amine (–NH₂) groups. How can the purity of the compound be determined?

12–9. An entomologist working on mosquito breeding has set up an artificial stream in his laboratory. Unfortunately the larvae keep dying. Metal poisoning from plumbing in the circulating water system is suspected. Among the metals that may be present are copper, iron, and zinc. How could each of these metals be determined at the part-per-million level?

12–10. Axel Grind was asked to devise a method for determining the ammonium ion in ammonium phosphate. He proposed an acid–base method as follows: "Treat a sample with excess sodium hydroxide and distill the released ammonia. Carefully collect the distillate in excess boric acid solution. Using methyl red indicator, titrate the ammonium borate–boric acid solution with a standard solution of hydrochloric acid." Is Axel's proposed method feasible?

12–11.   The bearings in a car motor have worn out prematurely. Leakage of ethylene glycol antifreeze into the crankcase is suspected to be the cause. Suggest a method for the determination of ethylene glycol in motor oil.

12–2. ANALYTICAL LITERATURE: A LITERATURE-SEARCH PROBLEM

> The more extensive a man's knowledge of what has been done, the greater will be his power of knowing what to do.
>
> *Disraeli*

Background

In Section 12-1 the point is made that analytical chemistry is not so much performing analyses as it is solving chemical problems. Problems of analysis are solved with the assistance of two main resources. First and indispensable is knowledge of and practical experience with techniques of sample preparation, separations, and measurement. Second is the accumulated knowledge inherited from predecessors who have worked on similar problems. The purpose of the experimental work in this book has been to build experience in the practice and applicability of a number of important methods of analysis. This experience, melded with pertinent information from the literature, should provide a solid foundation for attacking analytical problems.

The Literature of Analytical Chemistry

Presently about 10,000 papers in the area of analytical chemistry are published each year. Consulting this store of information often reveals that a specific analysis can be performed by any of several methods; the question is then one of intelligent selection and, if necessary, adaptation. A specialist is likely to attempt to solve the problems he encounters by the technique with which he is most familiar, be it mass spectrometry, polarography, chromatography, or any other. For one problem, mass spectrometry might be better than polarography, for another the reverse might be true, and for still another neither would be suitable. An intelligent decision as to the most appropriate method must be based on sufficient knowledge of the whole arsenal of methods available. Usually a published procedure cannot be used directly without some adaptation or development that also must be based on intelligent choices. Chemical literature includes the following:

1. **Abstracts.** *Chemical Abstracts,* the most important, is published by the American Chemical Society in 26 issues per year. Extensive subject and author indexes are provided. It contains short abstracts of the world's chemical literature from more than 12,000 journals and periodicals.

    Specialized abstracts are available in areas such as gas chromatography and nuclear magnetic resonance spectroscopy.

2. **Collections of Methods and Data.** These serve as sources of information about theoretical background, practical procedures, and fundamental data. Examples are Scott's *Standard Methods of Chemical Analysis, Treatise on Analytical Chemistry,* and *Encyclopedia of Industrial Chemical Analysis.*

3. **Books and Reviews on Specific Topics.** A selected list dealing with restricted areas of analytical chemistry is given by D. A. Skoog and D. M. West in *Fundamentals of Analytical Chemistry,* 2nd ed., Holt, Rinehart, and Winston, New York, 1969, p 722.

4. **Periodicals.** These normally contain reports of original research. The most widely read articles on analysis appear in *Analytical Chemistry,* published by the American Chemical Society. Other journals emphasizing analytical research include *Talanta, Analytica Chimica Acta, The Analyst, Zeitschrift für Analytische Chemie,* and *Journal of Analytical Chemistry* (USSR). Additional journals are listed by Skoog and West (see Item 3 above).

Searching the Chemical Literature

An examination of the most recent issues of *Chemical Abstracts* usually will provide several references to the original literature along with brief statements of the author's findings. At the beginning it is advisable not to spend much time in the older abstracts. The recent original literature will include references leading to virtually all the pertinent preceding work and, in addition, provide critical comment about preceding work that is usually lacking in abstracts. Although collections are likely to suggest only well tested methods, they tend to go out of date quickly and often are superseded by better methods in the current literature. Books are most useful in the later stages of decision making.

They provide in-depth discussions of the theory and practice of selected techniques and are generally more up-to-date than collections of methods. Scanning current issues of well chosen periodicals often will uncover the latest pertinent reports.

In this experiment an analytical problem is assigned to provide some experience in using chemical literature and in selecting the techniques most appropriate for a given situation. These problems are intended to require consultation of the literature, and it is not presumed that the answer will be immediately obvious. The proposed method need not be limited to the techniques introduced in this book.

Experimental Procedure

A variety of problems may be classified under areas such as chemical, biochemical, biological, environmental, industrial, geological, food, agricultural, and clinical. Choose a major area. A specific problem in that area will be assigned. Guided by the literature and your experience, propose a solution to the problem.

Submit a summary of the proposed solution in 150 to 200 words. Include (1) a statement of the problem; (2) the proposed solution, with comments about limitations, precision, and instrument requirements; (3) reasons (in a sentence or two) why an answer to this problem might be of interest and what its practical significance might be; (4) the most recent significant reference located; (5) the most helpful reference(s); and (6) the search technique found to be most fruitful.

Some examples of literature-search problems are the following (area is given in parentheses):

1. Describe a method for the determination of the organo-phosphorus pesticides malathion and parathion in a cereal grain. (Agricultural)
2. The concentration of sulfur dioxide present in the atmosphere has been used as an indicator of urban air pollution. What is a simple, rapid method for the determination of sulfur dioxide in air at the level of 0.01 part per million? (Environmental)
3. Among the chemicals considered hazardous to health in cigarette smoke is catechol. How can the amount of catechol in cigarette smoke be determined? (Biochemical)
4. Some illnesses give rise to unusual proportions of alpha, beta, and gamma globulins in blood serum. How can the ratio of these proteins be determined? (Clinical)
5. Insulating oils for electrical power transformers must be kept as water-free as possible if their insulating capacity is to remain high. How can the water content of such an oil be determined? (Industrial)

6.   Molybdenum enhances the hardness, strength, and corrosion resistance of steels. How can the amount of molybdenum in a tool steel be determined? (Industrial)

7.   How can the amount of secondary butyl bromide in isobutyl bromide be measured? (Chemical)

8.   The molecules responsible for the transport of oxygen in the blood of crabs and other marine arthropods contain copper, unlike those of mammals, which contain iron. How can the amount of copper in the blood of a crab be determined? (Biological)

9.   How can the phosphorus content of milk be determined? (Food)

10.   How can the polychlorobiphenyls in a sample of drinking water be determined? (Environmental)

11.   How could oxygen at the level of a part per thousand be determined in a gaseous mixture of nitrogen and carbon dioxide? (Chemical)

12.   How could the titanium in a sample of Australian beach sand containing rutile, zircon, garnet, and quartz be determined? (Geological)

SELECTED REFERENCES

C. R. Burman, *How to Find Out in Chemistry,* Pergamon Press, Elmsford, N.Y., 1966. A useful guide to sources of information.

E. I. Crane, A. M. Patterson, and E. B. Marr, *A Guide to the Literature of Chemistry,* 2nd ed., Wiley, New York, 1957.

M. G. Mellon, *Chemical Publications–Their Nature and Use,* 4th ed., McGraw-Hill, New York, 1965.

LABORATORY EQUIPMENT

This Appendix lists the equipment necessary for the performance of the experiments in this textbook. The items listed in Tables A-1 and A-2 are most conveniently provided in individual lockers. Items listed separately in Table A-3 are required for only a single experiment; these should be checked out as needed and returned after use. Major items of equipment provided in the laboratory for community use are given in Table A-4.

When checking in, make sure that all required items are present in perfect condition.

Before checking out, clean all equipment and return borrowed material. Dismantle wash bottles and discard the glass tubing. Discard stirring rods and used expendable items. Wash all equipment the period before so that it is clean and dry at the time of checkout.

TABLE A–1. SUGGESTED EXPENDABLE ITEMS IN INDIVIDUAL LOCKERS

Item	Approximate 1972 Unit Cost, $	Quantity
Laboratory towel	0.12	1
Matches	0.03	1
Medicine droppers	0.05	3
Pipe cleaner, 18 in.	0.03	1
Policemen	0.05	3
Rubber bulbs, 2 ml	0.06	3
Sponge	0.18	1

TABLE A–2. SUGGESTED APPARATUS FOR INDIVIDUAL
LOCKERS

Item	Approximate 1972 Unit Cost, $	Quantity
Beakers		
50 ml	0.41	8
150 ml, 250 ml	0.41	1 each
400 ml	0.48	6
600 ml	0.60	1
Bottle, 1 liter, polyethylene	0.65	1
Bottles, 1 liter, glass	0.25	2
Bulb, rubber, 25 ml	0.36	1
Bunsen burners	3.25	3
Buret clamps	2.05	3
Camel-hair brush	0.15	1
Conical flasks		
50 ml	0.58	1
200 ml	0.50	9
500 ml	0.50	2
Crucibles		
Gooch	1.19	3
Ordinary	0.50	3
Crucible tongs	1.75	1
Funnels		
Buret	0.40	1
Ordinary, 65 mm, 58°	0.68	3
Powder	0.95	1
Funnel support	2.00	1
Graduated cylinders, 10 ml, 100 ml	0.80	1 each
Planchets for radioactivity counting	0.15	9
Rings, 4 in.	0.95	3
Rubber finger cots	0.05	2
Rubber stopper, No. 1, one hole	0.05	2

(*Table A-2 continues on the opposite page*)

TABLE A−2. SUGGESTED APPARATUS FOR INDIVIDUAL
LOCKERS (*Continued*)

Item	Approximate 1972 Unit Cost, $	Quantity
Spatula	0.70	1
Test-tube brushes, large, small	0.12	1 each
Test tubes, Pyrex		
25 X 150 mm	0.16	6
13 X 100 mm	0.06	7
Triangles	0.30	3
Sample tubes, plastic	0.07	3
Triangular file	0.25	1
Wash bottles, polyethylene,		
250 ml, 500 ml	0.65	1 each
Watch glasses, 100 mm	0.19	4
Weighing bottles	1.35	2
Wing top	0.45	1
Wire gauzes	0.27	3

Check out when needed and store in lockers:

Buret		
Stopcock only, Teflon	4.50	1
Barrel only	4.50	1
Pipets		
10 ml	1.45	1
20 ml	1.70	1
Volumetric flasks		
50 ml	2.75	6
100 ml	3.00	2
250 ml	3.65	1
500 ml	4.25	1

TABLE A–3. SUGGESTED APPARATUS FOR SINGLE
EXPERIMENTS

Item	Experiment Number	Approximate 1972 Unit Cost, $	Quantity
Thermometer	1–5	2.22	1
Pipet, 100 ml	2–3	3.25	1
Pipet, measuring, 2 ml	6–2	2.20	1
Nessler tubes, 100 ml	6–2	2.10	5
Syringe, 10 ml	7–2	4.00	1
Glass electrode	8–2	25.00	1
Calomel reference electrode	8–2	20.00	1
Reference-electrode assembly			
Inner tube	8–3	1.65	1
Cap	8–3	2.50	1
Pt and Ag wires	8–3	4.50	1 each
Battery, 1.5 V	8–3	1.25	1
Voltmeter	8–3	60.00	1
Generating-electrode pair	8–4	20.00	1
Amperometric end-point detection system	8–4	10.00	1
Planimeter	9–2	140.00	1
Template	9–2	2.00	1
Pipet, 9 in. disposable	10–2, 10–3	0.05	1
Filter assembly for thorium-234	10–5	10.00	1
Cation-exchange columns	11–2	3.00	2
Anion-exchange columns	11–3, 11–4	3.00	2
Siphon pipet, 3 to 4 ml	11–4	0.50	1
Syringe for gas chromatography, ¼ ml	11–5	4.00	1

TABLE A–4. SUGGESTED EQUIPMENT IN LABORATORY FOR
COMMUNITY USE

Analytical balance, 0.1-mg sensitivity
Coulometer, constant current
Drying oven
Electronic calculator
First-aid kit
Gas chromatograph
Geiger counter
Hot plate
Large ion-exchange column for distilled-water purification
Nessler-tube rack
pH meter
Sodium bicarbonate for acid and base spills
Spectrophotometers, visible and atomic absorption
Suction flask, 2 liter, with vacuum-filtration assembly
Top-loading balance, 1-mg sensitivity
Triple-beam balance, 0.01-g sensitivity

CHEMICALS

Acids and Bases

6 M NH$_3$, HCl, HNO$_3$, acetic acid; 1 M NaOH; 3 M H$_2$SO$_4$; 12 M HCl. (NH$_3$ and 12 M HCl are to be kept in fume hood.)

Solutions Provided for Individual Experiments

1–2   Preliminary: 10% CoCl$_2$ solution.

1–5   Aliquot: 5% KMnO$_4$ solution.

3–1   Chloride: 5% K$_2$CrO$_4$ solution; 0.1% dichlorofluorescein in 70% alcohol; 0.5 M acetic acid–0.5 M sodium acetate buffer, pH 4.7 (add 20 g of NaOH per liter to 1 M acetic acid).

3–2   Carbonate: 0.1% bromocresol green solution.

3–3   Calcium: 1% MgCl$_2$ solution; 0.1% Calmagite solution; 0.7 M triethanolamine solution; 0.4 M sodium cyanide solution (*Caution:* poison).

4–2   Iron: saturated HgCl$_2$ solution; 6 M H$_3$PO$_4$; 0.5% sodium diphenylamine sulfonate indicator.

4–3   Iron: Zimmermann-Reinhardt solution (dissolve 70 g of MnSO$_4$·4H$_2$O in 500 ml of water, add 125 ml of 18 M H$_2$SO$_4$ and 125 ml of 85% H$_3$PO$_4$, dilute to 1 liter); saturated HgCl$_2$ solution (see 6-2 below for preparation).

4–4   Copper: fresh 4% urea solution.

5–2   Chloride: 0.1 M AgNO$_3$.

5–3   Nickel: 2% sodium dimethylglyoxime solution; 0.1% methyl red solution.

5–4   Sulfur: 0.05 M BaCl$_2$; 0.1 M AgNO$_3$.

6–2   Ammonia: Nessler's reagent. [Prepare the following solutions: (1) 200 g of KI in 300 ml of water; (2) 450 g of KOH in 1200 ml of water; (3) saturated $HgCl_2$, by addition of 70 g of $HgCl_2$ to 1000 ml of water and dissolution by heating and cooling. Slowly add the saturated $HgCl_2$ solution to the KI solution with constant stirring until a permanent red precipitate forms. Add solid KI (about 0.5 g) until the precipitate dissolves. Then add the entire KOH solution and dilute to about 2 liters. Store in a dark bottle.]

6–3   Iron: 0.1% bipyridine solution; 10% hydroxylamine hydrochloride solution; 10% sodium acetate solution.

6–7   Iron: 0.02% $MgCl_2$ solution; 0.2 M $CH_2ClCOOH$–0.05 M $CH_2ClCOONa$ buffer, pH 2.3 (add 2 g of NaOH to 24 g of monochloroacetic acid in methanol); 0.1% Calmagite solution; 6% salicylic acid solution in methanol.

7–2   Oxine: 0.1% 4-phenylazodiphenylamine solution.

8–3   Cerium: 18 M H_2SO_4.

10–2.   Trace nickel; 10% hydroxylamine hydrochloride solution; 1% dimethylglyoxime in ethanol.

10–3   Aluminum: 3 M hexamethylenetetramine solution; 0.2% Xylenol Orange indicator solution (prepare fresh weekly); saturated NaF solution.

10–4   Aluminum: 0.1% methyl red solution; 0.1% thymol blue solution.

10–5   Thorium-234: 0.03 M $FeCl_3$ in 0.1 M HCl; 0.5 M $(NH_4)_2CO_3$; standard solution of aged $UO_2(NO_3)_2$ (thorium-234).

11–2   Salt: 0.1% methyl red solution.

11–3   Zinc–nickel: 0.2% murexide indicator (solid, ground with NaCl); 0.1% Calmagite indicator solution.

11–4   Chloride–bromide: 5% K_2CrO_4 solution.

Pure Chemicals and Materials Required

3–1   Chloride: $NaHCO_3$; $AgNO_3$; dextrin; $CaCO_3$.

3–2   Carbonate: Na_2CO_3.

3–3   Calcium: EDTA; $CaCO_3$; NH_4Cl.

4–2   Iron: $K_2Cr_2O_7$; electrolytic Fe; $SnCl_2 \cdot 2H_2O$; $KMnO_4$.

4–3   Iron: $SnCl_2 \cdot 2H_2O$; $KMnO_4$; $Na_2C_2O_4$.

4–4   Copper: $Na_2S_2O_3 \cdot 5H_2O$; Na_2CO_3; copper wire cut into 0.2-g portions; KI; starch; NH_4HF_2; chloroform.

4–5   Glycol: As_2O_3; $NaHCO_3$; KI; H_5IO_6; I_2; starch.

5–2   Chloride: borosilicate-glass fiber mats.

5–3   Nickel: borosilicate-glass fiber mats; tartaric acid.

5–4   Sulfate: borosilicate-glass fiber mats.

6–3   Trace iron: $FeSO_4 \cdot (NH_4)_2SO_4 \cdot 6H_2O$; $NH_2OH \cdot HCl$.

6–4   Copper: copper wire; acetylene.

6–5   Cobalt–nickel: cobalt powder; nickel powder; EDTA.

6–7   Iron: EDTA; $CaCO_3$; NH_4Cl.

7–2   Oxine: *p*-toluenesulfonic acid; acetonitrile.

8–2   Acid mixture: sulfamic acid; NaOH.

8–3   Cerium: $K_4Fe(CN)_6 \cdot 3H_2O$; NaCl; Na_2CO_3.

8–4   Cyclohexene: KBr; glacial acetic acid; methanol; mercury(II) acetate.

10–2   Trace nickel: chloroform; nickel powder; sodium acetate; sodium hydrogen tartrate; sodium thiosulfate; 10-cm qualitative medium-porosity filter paper.

10–3   Aluminum: zinc metal; EDTA; methyl isobutyl ketone.

10–4   Aluminum: succinic acid; urea; methyl isobutyl ketone; medium-porosity low-ash 10-cm filter paper.

11–2   Salt: strong-acid cation resin, 50 to 100 mesh, 8% cross-linked; NaOH; KCl; paraffin.

11–3   Zinc–nickel: EDTA; strong-base anion-exchange resin, 50 to 100 mesh, 8% cross-linked; paraffin; NH_4Cl.

11–4   Chloride–bromide: $NaNO_3$ (chloride free); $NaHCO_3$; $AgNO_3$; strong-base anion-exchange resin, 50 to 100 mesh, 8% cross-linked; paraffin.

11–5   Cyclohexane: helium; benzene; cyclohexane.

Appendix C

TABLES OF DATA

TABLE C–1. IONIZATION CONSTANTS

A. Acids

Acetic		1.9×10^{-5}	Phenol		1	$\times 10^{-10}$
Benzoic		6.7×10^{-5}	Phosphoric	K_1	7	$\times 10^{-3}$
Boric	K_1	5×10^{-10}		K_2	7	$\times 10^{-8}$
Carbonic	K_1	3.5×10^{-7}		K_3	4	$\times 10^{-13}$
	K_2	5×10^{-11}	Phthalic	K_2	4	$\times 10^{-6}$
Chromic	K_2	1×10^{-7}	Silicic	K_1	2	$\times 10^{-10}$
Citric	K_1	8.7×10^{-4}	Succinic	K_1	6	$\times 10^{-5}$
	K_2	1.8×10^{-5}		K_2	2	$\times 10^{-6}$
	K_3	4×10^{-6}	Sulfuric	K_2	1.2	$\times 10^{-2}$
Formic		2×10^{-4}	Sulfurous	K_1	2	$\times 10^{-2}$
Hydrocyanic		7×10^{-10}		K_2	6	$\times 10^{-8}$
Hydrofluoric		7×10^{-4}	EDTA	K_1	7	$\times 10^{-3}$
H_2S	K_1	9×10^{-8}		K_2	2	$\times 10^{-3}$
	K_2	1×10^{-15}		K_3	7	$\times 10^{-7}$
Oxalic	K_1	6×10^{-2}		K_4	6	$\times 10^{-11}$
	K_2	6×10^{-5}				

B. Bases

Ammonia	1.8×10^{-5}	Monoethanolamine	3	$\times 10^{-5}$
Hydroxylamine	1×10^{-8}	Pyridine	1	$\times 10^{-9}$
Methylamine	4×10^{-4}	2-Aminopyridine	5	$\times 10^{-8}$
$CH_2 ClCOOH$	1.5×10^{-3}			

C. Ion product for water at 24°C 1×10^{-14}

TABLE C–2. SOLUBILITY PRODUCTS

Aluminum hydroxide	2×10^{-32}
Barium carbonate	5×10^{-9}
Barium chromate	1×10^{-10}
Barium oxalate	2×10^{-8}
Barium sulfate	1×10^{-10}
Cadmium sulfide	1×10^{-28}
Calcium carbonate	5×10^{-9}
Calcium fluoride	4×10^{-11}
Calcium oxalate	2×10^{-9}
Copper(II) hydroxide	2×10^{-20}
Copper(II) sulfide	1×10^{-36}
Iron(III) hydroxide	1×10^{-36}
Lead chromate	2×10^{-14}
Lead sulfate	2×10^{-8}
Lead sulfide	1×10^{-28}
Magnesium carbonate	1×10^{-5}
Magnesium hydroxide	1×10^{-11}
Magnesium ammonium phosphate	2×10^{-13}
Magnesium oxalate	9×10^{-5}
Manganese(II) sulfide	1×10^{-15}
Mercury(I) bromide	3×10^{-23}
Mercury(I) chloride	6×10^{-19}
Mercury(II) sulfide	1×10^{-52}
Potassium perchlorate	2×10^{-2}
Silver argenticyanide	4×10^{-12}
Silver bromide	8×10^{-13}
Silver carbonate	6×10^{-12}
Silver chloride	1×10^{-10}
Silver chromate	2×10^{-12}
Silver iodide	1×10^{-16}
Silver phosphate	1×10^{-19}
Silver sulfide	1×10^{-50}
Silver thiocyanate	1×10^{-12}
Strontium chromate	4×10^{-5}
Zinc hydroxide	5×10^{-18}
Zinc sulfide	1×10^{-24}

TABLE C–3. FORMATION CONSTANTS

$Ag(NH_3)_2^+$	1.4×10^7
$Zn(NH_3)_4^{++}$	3×10^9
$Cu(NH_3)_4^{++}$	1×10^{13}
$Cd(NH_3)_4^{++}$	1×10^7
$Ni(NH_3)_4^{++}$	1×10^8
$Co(NH_3)_6^{3+}$	2.5×10^{34}
$Fe(CNS)^{++}$	1.4×10^2
$Hg(CNS)_4^=$	5×10^{21}
$Ag(CN)_2^-$	5×10^{20}
$Zn(CN)_4^=$	5×10^{16}
$Cu(CN)_4^{3-}$	5×10^{27}
$Fe(CN)_6^{4-}$	1×10^{24}
$Fe(CN)_6^{3-}$	1×10^{31}
$Cd(CN)_4^=$	2×10^{18}
$Hg(CN)_4^=$	1×10^{42}
$HgCl_4^=$	1×10^{16}
$HgI_4^=$	2×10^{30}
$Ag(S_2O_3)_2^{3-}$	2×10^{13}
$Ca(EDTA)^=$	5×10^{10}
$Mg(EDTA)^=$	5×10^8
$Al(OH)_4^-$	2×10^{32}
$Cr(OH)_4^-$	1×10^{32}
$Pb(OH)_3^-$	1×10^{14}
$Zn(OH)_4^=$	2×10^{15}

TABLE C–4. ELECTRODE POTENTIALS[a]

Couple	E^0
$Na^+ + e = Na$	-2.71
$Mg^{++} + 2e = Mg$	-2.37
$Al^{3+} + 3e = Al$	-1.66
$2H_2O + 2e = H_2 + 2OH^-$	-0.83
$Zn^{++} + 2e = Zn$	-0.76
$Fe^{++} + 2e = Fe$	-0.44
$Cr^{3+} + e = Cr^{++}$	-0.41
$Cd^{++} + 2e = Cd$	-0.40
$Tl^+ + e = Tl$	-0.34
$V^{3+} + e = V^{++}$	-0.26
$Sn^{++} + 2e = Sn$	-0.14
$Pb^{++} + 2e = Pb$	-0.13
$2H^+ + 2e = H_2$	0.00
$TiO^{++} + 2H^+ + e = Ti^{3+} + H_2O$	0.10
$S + 2H^+ + 2e = H_2S$	0.14
$Sn^{4+} + 2e = Sn^{++}$	0.15
$Cu^{++} + e = Cu^+$	0.15
$S_4O_6^= + 2e = 2S_2O_3^=$	0.17
$SO_4^= + 4H^+ + 2e = H_2O + H_2SO_3$	0.17
$AgCl + e = Cl^- + Ag$	0.22
Saturated calomel	$(E = 0.24)$
$Hg_2Cl_2 + 2e = 2Cl^- + 2Hg$	0.27
$Cu^{++} + 2e = Cu$	0.34
$Fe(CN)_6^{3-} + e = Fe(CN)_6^{4-}$	0.36
$Cu^+ + e = Cu$	0.52
$I_2 + 2e = 2I^-$	0.53
$H_3AsO_4 + 2H^+ + 2e = H_3AsO_3 + H_2O$	0.56
$2HgCl_2 + 2e = Hg_2Cl_2 + 2Cl^-$	0.63
$O_2 + 2H^+ + 2e = H_2O_2$	0.68
Quinone $+ 2H^+ + 2e =$ hydroquinone	0.70
$Fe^{3+} + e = Fe^{++}$	0.77
$Hg_2^{++} + 2e = 2Hg$	0.79
$Ag^+ + e = Ag$	0.80
$Hg^{++} + 2e = Hg$	0.85
$NO_3^- + 3H^+ + 2e = HNO_2 + H_2O$	0.94
$HNO_2 + H^+ + e = NO + H_2O$	1.00
$VO_2^+ + 2H^+ + e = VO^{++} + H_2O$	1.00
$Br_2(l) + 2e = 2Br^-$	1.06
$2IO_3^- + 12H^+ + 10e = 6H_2O + I_2$	1.20
$O_2 + 4H^+ + 4e = 2H_2O$	1.23
$MnO_2 + 4H^+ + 2e = Mn^{++} + 2H_2O$	1.23
$Cr_2O_7^= + 14H^+ + 6e = 7H_2O + 2Cr^{3+}$	1.33
$Cl_2 + 2e = 2Cl^-$	1.36
$BrO_3^- + 6H^+ + 6e = 3H_2O + Br^-$	1.44
$MnO_4^- + 8H^+ + 5e = 4H_2O + Mn^{++}$	1.51
$Ce^{4+} + e = Ce^{3+}$ ($1M$ HNO_3)	1.61

[a] Data at $25°C$ and under standard conditions.

TABLE C–5. ACID–BASE INDICATORS AT 25°C

Indicator	Transition Range	pK_{in}	Acid	Base
Thymol blue	1.2– 2.8	1.6	Red	Yellow
Methyl yellow	2.9– 4.0	3.3	Red	Yellow
Methyl orange	3.1– 4.4	4.2	Red	Yellow
Bromocresol green	3.8– 5.4	4.7	Yellow	Blue
Methyl red	4.2– 6.2	5.0	Red	Yellow
Chlorophenol red	4.8– 6.4	6.0	Yellow	Red
Bromothymol blue	6.0– 7.6	7.1	Yellow	Blue
Phenol red	6.4– 8.2	7.4	Yellow	Red
Cresol purple	7.4– 9.0	8.3	Yellow	Purple
Thymol blue	8.0– 9.6	8.9	Yellow	Blue
Phenolphthalein	8.0– 9.8	9.7	Colorless	Red
Thymolphthalein	9.3–10.5	9.9	Colorless	Blue

TABLE C–6. CONCENTRATED ACIDS AND BASES

	Molecular Weight	Density	Weight percent	Molarity
CH_3COOH	60.05	1.05	99.5	17.4
H_2SO_4	98.07	1.83	94	17.6
HF	20.01	1.14	45	25.7
HCl	36.46	1.19	38	12.4
HBr	80.91	1.52	48	9.0
HNO_3	63.01	1.41	69	15.4
$HClO_4$	100.46	1.67	70	11.6
H_3PO_4	98.00	1.69	85	14.7
NaOH	40.00	1.53	50	19.1
NH_3	17.03	0.90	28	14.8

TABLE C–7. FORMULA WEIGHTS OF SOME COMMON RADICALS

AsO_3	122.92	CrO_4	115.99	NO_3	62.00
AsO_4	138.92	Cr_2O_7	215.99	OH	17.007
BrO_3	127.90	$Fe(CN)_6$	211.96	PO_4	94.97
CO_3	60.01	IO	142.90	P_2O_7	173.94
C_2O_4	88.02	IO_2	158.90	$PtCl_6$	407.81
CN	26.018	IO_3	174.90	SO_3	80.06
$C_2H_3O_2$		IO_4	190.90	SO_4	96.06
(acetate)	59.04	MnO_4	118.94	S_2O_3	112.12
ClO	51.45	MoO_4	159.94	SCN	58.08
ClO_2	67.45	NH_4	18.039	SiO_3	76.08
ClO_3	83.45	NO_2	46.006	UO_2	270.03
ClO_4	99.45				

TABLE C–8. FORMULA WEIGHTS OF SOME COMMON COMPOUNDS[a]

$AgBr$	187.77	$HC_7H_5O_3$	138.12	Na_2CO_3	105.99
$AgCl$	143.32	(salicylic acid)		$NaC_2H_3O_2$	82.03
Ag_2CrO_4	331.73	H_2O	18.015	$Na_2C_2O_4$	134.00
AgI	234.77	H_2O_2	34.015	$NaHCO_3$	84.01
$AgNO_3$	169.87	$HgCl_2$	271.50	$NaCl$	58.44
Al_2O_3	101.96	HgO	216.59	$NaClO_4$	122.44
As_2O_3	197.84	I_2	253.81	$Na_2H_2Y\cdot2H_2O$	372.24
$BaCO_3$	197.35	KBr	119.01	(Y = EDTA anion)	
$BaCl_2$	208.25	K_2CO_3	138.21	$NaNO_3$	84.99
$BaSO_4$	233.40	KCl	74.56	$NaOH$	39.997
CO_2	44.010	$KClO_4$	138.55	Na_3PO_4	163.94
$CaCO_3$	100.09	K_2CrO_4	194.20	Na_2SO_4	142.04
CaC_2O_4	128.10	$K_2Cr_2O_7$	294.19	$Na_2S_2O_3$	158.10
$CaCl_2$	110.99	$KHC_8H_4O_4$	204.23	NH_3	17.031
CaO	56.08	(KH phthalate)		NH_4Cl	53.49
$CaSO_4$	136.14	$K_3Fe(CN)_6$	329.26	$(NH_4)_2Ce(NO_3)_6$	548.23
CuI_2	317.36	$K_4Fe(CN)_6\cdot3H_2O$	422.41	$(NH_4)_2CO_3$	96.09
$CuSO_4$	159.60	KI	166.01	$(NH_4)_2SO_4$	132.14
Ethylene glycol	62.07	KIO_3	214.00	$PbCl_2$	278.1
$Fe(NH_4)_2(SO_4)_2$		$KMnO_4$	158.04	$Pb(NO_3)_2$	331.2
$6H_2O$	392.13	KNO_3	101.11	PbO_2	239.2
Fe_2O_3	159.69	KOH	56.11	$PbSO_4$	303.3
Fe_3O_4	231.54	$KSCN$	97.18	SO_2	64.06
$HCOOH$	46.026	K_2SO_4	174.26	SiO_2	60.08
(formic acid)		$LiBr$	86.84	$SnCl_2$	189.60
H_2CO_3	62.03	LiI	133.85	SnO_2	150.69
$HC_2H_3O_2$	60.05	$MgCO_3$	84.31	ThO_2	264.04
(acetic acid)		$MgCl_2$	95.21	TiO_2	79.90
$H_2C_2O_4\cdot2H_2O$	126.07	$Mg(NO_3)_2$	148.31	U_3O_8	842.08
(oxalic acid)		$MgSO_4$	120.36	V_2O_5	181.88
$HC_7H_5O_2$	122.12	MnO_2	86.94	WO_3	231.85
(benzoic acid)		Na_3AsO_3	191.89	$Zn(NO_3)_2$	189.38
HNH_2SO_3	97.09	$NaBr$	102.89	$ZnSO_4$	161.43
(sulfamic acid)					

[a] For common inorganic acids see Table C-6.

Appendix D

ANSWERS TO SELECTED PROBLEMS

In some cases the uncertainty in the last digit of the answers given here may be more than 1. Normally a relative precision of 1 part per 1000 is assumed.

Figure 1-7, Vernier readings: 40.6048, 7.0430, 24.1763, −0.0006.

Chapter 1.

1–1 No

1–3 (Using 0.0011 as density of air) 20.0053; 19.9982; 20.0000; 20.0053 g

1–4 1,2,4,18,16,32, and 64 g; 1,2,2,5,10,20, 20, and 50 g

1–5 Up, 5 revolutions

1–6 (Using 0.0011 as density of air) 5.0018 g

1–7 Because the full-scale reading is set *vs* a stainless-steel wt

1–8 25.016 ml

1–9 To ± 0.025 g

1–10 High, by 0.46%

1–11 More, because of slower drainage

1–12 0.002 ml, or 0.02%; drainage

1–14 0.2001; 0.1999; 0.1996 M; 4°C

1–15 Transfer to a 1-liter flask

1–16 Low; either high or low; high

1–17 0.001138; 0.001134; 0.001136

Chapter 2.

2–1 1,2,4,70

2–2 43.64; 0.7778; 4.426 × 10^{-7}; 3.000 × 10^4

2–3	(a) 0.06, 0.002; (b) 0.002, 0.003; (c) 5, 0.004	2–4	231.735; 69.62; 258.32; 239.3; 18.015
2–5	40.07; 0.06; 40.06 (40.32 rejected by Q test); 0.05; 0.06; 40.06 ± 0.06	2–6	(a) 12.37; 0.02; 12.36 (12.45 rejected by Q test); 0.02; 0.02; 12.36 ± 0.02; (b) 0.432; 0.002; 0.432; 0.002, 0.003, 0.432 ± 0.005; (c) 122.2, 0.4, 122.4 (120.2 rejected by Q test), 0.3, 0.4, 122.4 ± 0.6; (d) 87.40, 0.12; 87.44, 0.16, 0.23, 87.44 ± 0.36

Chapter 3.

3–1	$47.3_4\%$	3–2	$26.9_2\%$
3–3	18.42 ml	3–4	0.582_1 g
3–5	(a) 0.072, 0.0073, 1.07 × 10^{-6}, 7.6 × 10^{-7}; 1.14, 2.14, 7.97, 8.12; (b) 1 × 10^{-5}, 5	3–6	$2\,M$
3–7	0.41 to 0.44 g	3–8	373 liters
3–9	$0.185_3\,M$	3–10	$50.1_6\%$
3–11	$NaOH$ and Na_2CO_3, $NaHCO_3$ and Na_2CO_3, $NaHCO_3$ and H_2CO_3	3–13	$NaOH$ and Na_2CO_3, 58.2_5 and $32.3_9\%$
3–14.	11.13, 9.26, 5.28; methyl red	3–16	Yes
3–17	0.96, 0.20	3–18	50 ml
3–19	$85.6_6\%$	3–20	74.7; 74.7; 62.9
3–21	89 ml	3–23	$10^{18.1}$

Chapter 4.

4–1	$0.01691\,M$	4–2	$0.01786\,M$
4–3	49.04%		
4–4	$59.1_0\%$	4–5	9.88%
4–6	$58.2_5\%$	4–8	$54.0_5\%$
4–9	$5 \times 10^{-13}\,M$	4–10	$63.5_4\%$
4–11	$18.4_3\%$	4–12	$0.0501\,M$
4–13	$1.4 \times 10^{-6}\,M$	4–14	$52.0_2\%$
4–15	$0.1243\,M$	4–16	1.4×10^{-13}
4–18	$0.0622_4\,M$	4–19	0.877_1 g
4–20	No	4–21	0.44 V

Chapter 5.

5–1	$66.0_8\%$	5–2	60 ml
5–3	0.068%; 0.21%; 294 ml	5–4	11.84%
5–5	$65.4_6\%$	5–6	$23.7_6\%$; $97.9_3\%$
5–7	$83.9_9\%$	5–8	21.93%
5–9	123 ml	5–10	50 ml; 0.53 g

Chapter 6.

6–1	$3.2\ \mu g/ml$	6–2	$17\ \mu g$; $0.17\ \mu g/ml$
6–3	$0.668\ \mu g/ml$	6–4	$2.0\ \mu g/g$
6–5	2.04 mg	6–6	0.180; 30; $0.012\ M$
6–7	3.5×10^{-4}; 790	6–8	0.664 mg
6–9	1160	6–12	0.81%
6–13	$4.33 \times 10^{-4}\%$	6–15	19.6%; 20.2%
6–16	$5.00\ \mu g/ml$ Mn; $198\ \mu g/ml$ Cr.	6–18	At $0.1\ M$ and 650 nm; A = 0.622 in H_2O, 0.198 at pH 2, and 10.0 at pH 4.
6–21	$0.01463\ M$	6–22	$65.8_3\%$
6–25	6×10^{20}		

Chapter 7.

7–1	95.8%

Chapter 8.

8–1	0.00237_3; 0.001613	8–4	14.10%
8–5	14.86%	8–6	$65.5_2\%$
8–7	0.18 V	8–8	(a) undefined; (b) 0.36 V; (c) 0.98 V; (d) 1.61 V
8–9	−0.33 V; −0.35 V; −0.36 V; $-\infty$	8–10	+0.61 V
8–11	1.1×10^{-5}	8–12	26.4_7 mg

Chapter 9.

9–1	16.8; 0.05	9–2	Die 1: 15, 0.01; Die 2: 3.9, 0.6
9–3	3%; 6%		

Chapter 10.

10–1   1.58%

10–5   1.803 g

10–7   2; 3

10–9   3.625 pg

10–11 0.0054%

10–14 43.5 days

10–3   1.313 g

10–6   18.14%

10–8   4.6; 5.6

10–10 9.74×10^{-16} g

10–13 2.81×10^{-3}

10–15 8900 yr

Chapter 11.

11–1 3.98_3 meq; 15.8_6 meq/g

11–4   NaCl, 87.9_7%; $NaNO_3$, 12.03%

11–6 34.6_1%; 35.5_6%

11–8   HCl and HNO_3, then NaCl, then $CaCl_2$

11–11 53.5%

11–13 1.7; 73; 190

11–2   50.02%

11–5   (a) 9.89; (b) 0; (c) 11.51; (d) 6.10; (e) 0; (f) indeterminate

11–7   24.29%

11–10 21.51% Zn; 10.94% Cd

11–12 4.0; 1.53

11–15 About 100

INDEX

In this Index, page numbers in italic type indicate illustrations. Page numbers followed by (t) indicate tables.

Absolute error, 37, 38
Absorbance, 118
  maxima, location of, 148
Absorption
  nomenclature pertaining to, 118
  of CO_2, 92
  of light, 117
  of radiant energy, 117, 119, 132
  of water, 92
  quantitative aspects of light, 118
Absorption spectrophotometry. *See* Spectrophotometry.
Absorptivity, 118
Abstracts, Chemical, 252
Accidents, 1
Accuracy. *See also* Deviation; Errors; Precision.
  definition of, 35
  in pH measurements, 172
  in spectrophotometry, 126
Acetic acid, as solvent for acid-base titrations, 159
Acetonitrile, 158
  thermal expansion of, 160
  volatility of, 160
Acid(s). *See also* name of acid.
  composition and density of concentrated, 266(t)
  dissociation constants of, by potentiometry, 168, 262(t)
  ionization constants of, 262(t)
  potentiometric titration of, 170
  primary standard, 61
  strengths of, solvent effect on, 158
Acid-base indicators, 266(t)
Acid-base titration(s), 56
  of carbonate, titration curve for, 57
  primary standards for, 61
Adsorption
  and coprecipitation, 99, 203
  by precipitates, 99, 203
Adsorption indicators, 51
Aging, of precipitates, 100
Air, density of, 14

Air buoyancy, in weighing, 14
Aliquots
  checking technique for, 28
  disadvantage in use of, 50
  taking of, 28
Alkenes, determination of by bromination, 184
Alternation, 71
Aluminum
  EDTA complex with, 207, 208
  EDTA titration of, 206
  fluoride complex with, 208
  gravimetric determination of, 213
  precipitation of, 202, 213
  separation of, from iron(III), 202
Ammonia
  colorimetric determination of traces of, 119
  removal of, from water, 121
Amperometry, 183
Analytical balance, single pan, 5
  analytical weights in, 6, 7. *See also* Weights.
  beam of, 6, 7
  constant load, advantages of, 7
    design of, 6
    mechanism of, diagram of, *6, 7*
  direct-deflection principle of, 6, 7
  knife-edges of, 6, 7
  operating instructions for, 9
  optical-scale, sensitivity of, adjustment of, 11, 12, 13
    test of, 10
  principles of, 5
  reproducibility of, 10
  sensitivity of, 6, 11
  zero of, 9, 11, 12
  two-pan, 6
Analytical literature, 251
Analytical methods
  classifications of, *199*
  selection of, 248
Analytical results
  defective, 45
  reporting of, 43
  selection of best value in, 41

273